Student Solutions Manual
for
PHYSICS:
Calculus

Eugene Hecht
Adelphi University

Brooks/Cole Publishing Company

I(T)P® An International Thomson Publishing Company

Pacific Grove • Albany • Bonn • Boston • Cincinnati • Detroit
London • Madrid • Melbourne • Mexico City • New York • Paris
San Francisco • Singapore • Tokyo • Toronto • Washington

Assistant Editor: *Elizabeth Barelli Rammel*
Cover Design: *Vernon T. Boes*
Cover Photo: *Tom Skrivan*
Director Marketing Communications: *Margaret Parks*
Editorial Associate: *Beth Wilbur*
Production Editor: *Mary Vezilich*
Printing and Binding: *Malloy Lithographing*

COPYRIGHT© 1996 by Brooks/Cole Publishing Company
A division of International Thomson Publishing Inc.

I(T)P The ITP logo is a registered trademark under license.

For more information, contact:

BROOKS/COLE PUBLISHING COMPANY
511 Forest Lodge Road
Pacific Grove, CA 93950
USA

International Thomson Editores
Campos Eliseos 385, Piso 7
Col. Polanco
11560 México D. F. México

International Thomson Publishing Europe
Berkshire House 168-173
High Holborn
London WC1V 7AA
England

International Thomson Publishing GmbH
Königswinterer Strasse 418
53227 Bonn
Germany

Thomas Nelson Australia
102 Dodds Street
South Melbourne, 3205
Victoria, Australia

International Thomson Publishing Asia
221 Henderson Road
#05-10 Henderson Building
Singapore 0315

Nelson Canada
1120 Birchmount Road
Scarborough, Ontario
Canada M1K 5G4

International Thomson Publishing Japan
Hirakawacho Kyowa Building, 3F
2-2-1 Hirakawacho
Chiyoda-ku, Tokyo 102
Japan

Printed in the United States of America

5   4   3   2

ISBN 0-534-33986-7

Table of Contents

Preface

This Student Solutions Manual was written to complement *Physics: Calculus* by Eugene Hecht. It is designed to assist you in working independently to master the material discussed in the text.

Every chapter in the main text contains approximately twenty Discussion Questions. A selection of the odd-numbered questions are listed with italic numerals, and the answers to those questions appear in this manual. The Discussion Questions are meant to help you explore and develop your conceptual comprehension of the material. If an answer is not apparent, reread the appropriate portions of the text. The answers provided in this manual should be consulted only as a last resort or, alternatively, as a check.

Every chapter in the main text also contains approximately twenty Multiple Choice Questions. These are designed to test and develop your ability to deal quickly and accurately with conceptual issues that require little or no mathematical processing. Read the questions carefully. Make no assumptions about what the author might have had in mind. Respond only to the question as stated. This manual contains answers to the odd-numbered Multiple Choice Questions which are also answered in the main text.

Every chapter in the main text also contains about eighty numerical Problems listed according to increasing difficulty. A representative selection of the odd-numbered problems have italic numerals. Complete solutions for all of these problems appear in this manual. Again you are advised to attempt to solve each problem before consulting the Student Solutions Manual. The Answer Section of the main text contains a selection of skeletal solution. Study the solutions for similar problems before turning to the complete solutions provided in this manual.

Many people contributed to this project and deserve recognition, particularly Beth Wilbur, Editorial Associate, and Harvey Pantzis, Executive Editor.

Suggestions, corrections, and criticisms should be sent to Professor Gene Hecht, Physics Dept., Adelphi University, Garden City, NY 11530. Good luck!

Answers to Discussion Questions

1.1   Because on rare occasions there have been scientists who have simply lied about their work. The biologist who, in his laboratory, colored the butt of a white mouse with a black magic marker and then claimed to have performed a skin graft wasn't doing science. There are also a few cases of honest self-delusion in which what scientists have done has been utter nonsense and certainly not science.

1.3   Laws transcend experience because they generalize. We know that something has happened the same way for years and from that we conclude that it will continue to happen that way. Laws are timeless, and that goes beyond experience. The same can be said about the application of law everywhere — it would be strange if certain physical laws only worked in Chicago. Still, we certainly don't test any of them everywhere.

1.5   Science seeks to understand the Universe as we perceive it to be. This is to be done as best we can, knowing the tentative nature of theory. At the same time, we must be economical in our formulation. As William of Ockham (ca. 1290–1349) suggested: the simplest, briefest, least complex explanation is the one to be accepted.

1.7   Data are the uninterpreted perceptions (as much as that's possible). Facts are the product of interpretation within the context of some world view, some theoretical understanding. "I see a disc of light there" becomes "I see the sun god in heaven" or "I see a star in space." These are statements of fact based on different world views. Since world views can be biased, facts can be wrong.

1.9   The ultimate testing ground in physics is nature itself. If the universe is found experimentally to match the predictions of the theory, then it was "crazy enough."

Answers to Multiple Choice Questions

  1. a   3. a   5. c   7. d   9. d   11. d   13. d   15. b
17. a   19. c

Solutions to Problems

1.1 $10\,000\,000\,000 = 10^{10}$.

1.5 $1\ \text{m} = 1000\ \text{mm}$; $1\ \text{km} = 1000\ \text{m} = 1.00 \times 10^6\ \text{mm}$; $10.0\ \text{km} = 1.00 \times 10^7\ \text{mm}$.

1.9 $1\ \text{Å} = 10^{-10}\ \text{m}$; $5 \times 10^3\ \text{Å} = 5 \times 10^{-7}\ \text{m} = 5 \times 10^2\ \text{nm}$.

1.19 $65\ \text{cm}^2 = 10\ \text{in.}^2$; the dog's sensory area is $(10\ \text{in.}^2)/(3/4\ \text{in.}^2) = 13$ times greater.

1.23   299 792 458 m/s → 3×10^8 m/s; 3.00×10^8 m/s; 2.998×10^8 m/s; $2.997\,924\,6 \times 10^8$ m/s.

1.35   2.540 cm = 1.000 in.; 1 = (1.000 in.)/(2.540 cm) = 0.393 7 in./cm.

1.43 $(6 \times 10^5$ particles/h)(24 h/d)(365 1/4 d/y) = $5.259\,6 \times 10^9$ particles per year; total mass is (1.5 lb)(0.4536 kg/lb) = 0.6804 kg; hence (0.6804 kg)/(5.2596×10^9) = 1.2936×10^{-10} kg = 1×10^{-7} g. 50(0.68 kg) = 34 kg = 0.3×10^2 kg.

1.47   1.00 g + 0.001 00 g + 1.00×10^3 g + 0.000 001 00 g = 1001.001 001 g = 1.00 kg.

1.57

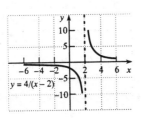

Answers to Discussion Questions

2.1   If the average speed is zero, the object didn't move, so it is not possible for it to have a nonzero speed over a smaller interval. The average velocity may well be zero even for a moving body if it returned to its original position. Hence, it could have a nonzero value over a shorter interval.

2.5   Yes. Imagine two people starting out at 9:00 a.m. on the same day, one at the top going down, the other at the bottom going up. They must meet at the same time somewhere.

2.7   No. A constant velocity means a constant speed. Yes. A constant speed can occur in a changing direction.

2.9   The dog is already running when the clock starts at $t = 0$ and it continues at a constant speed until $t = 2$ s, at which point it stops, having gone 3 m. It rests until $t = 4$ s and then runs at a constant speed until $t = 10$ s, whereupon it puts on a burst of speed. Nothing can be said about the displacement — the dog could be running in circles for all we know. It traveled a total distance of 6 m. It moved the fastest from 10 s to 11 s.

2.11   Changing the units will change the numerical value of the magnitude of a vector, but not its direction in space.

2.13   The mouse began moving at $t = 1$ s and $s_x = 1$ m. It ran to a point 2 m from the origin in 1 s and stopped. It rested for 1 s and then, at $t = 2$ s, it started running at nonuniform speed until it got 6 m from the origin. There, it immediately turned around and ran at a constant speed back to the opening of the tunnel.

2.17   (a) At $t = 0$, $s = 0/DC = 0$. (b) $s \approx (At^2 + Bt)/Dt$. (c) $s \approx (At^2 + Bt)/DC$.

Answers to Multiple Choice Questions

   1. d   3. b   5. c   7. c   9. c   11. d   13. b   15. a   17. c

Solutions to Problems

2.3 $v = l/t$; (a) $t = l/v = (1.0 \text{ ft})(0.304 \text{ 8 m/ft})/(2.998 \times 10^8 \text{ m/s}) = 1.0 \times 10^{-9}$ s. (b) $t = l/v = (1000 \text{ m})/(2.998 \times 10^8 \text{ m/s}) = 3.336 \times 10^{-6}$ s .

2.11 $\Delta t = 2.83 \text{ s} - 1.33 \text{ s} = 1.50 \text{ s}$; $l = vt = (10 \text{ m/s})(1.50 \text{ s}) = 15$ m.

2.21 $A = R^2$; $dA/dt = -2\pi R \, dR/dt = -2\pi Rv$

2.33 $V = (4/3)\pi r^3$; $dV/dt = 4\pi r^2 (dr/dt)$; $r = vt$; $dV/dt = 4\pi v^3 t^2 = 0.085$ m^3/s.

2.39 $100 = x^2 + y^2$;   (a) $dy/dt = - (x/y)dx/dt$;   (b) $dy/dt = - (1.5/9.89)1.0$ m/s $=$ $- 0.15$ m/s.   (c) It speeds up as it descends.

2.41 $s = \sqrt{(1.0 \text{ m})^2 + (1.0 \text{ m})^2} = 1.4$ m.

2.55 $s = \sqrt{(20 \text{ m})^2 + (60 \text{ m})^2} = 63$ m; $\tan\theta = 60/20$. $\theta = 71.6°$. s $= 63$ m $- 72°$ up from the horizontal.

2.69 $l_c = (10.0 \text{ m/s})t$ and $lv = (20.0 \text{ m/s})t$; $s(t) = \sqrt{l_c^2 + l_v^2 + (10.0 \text{ m})^2}$; $\dfrac{ds}{dt} =$

$$\frac{(500.0 \text{ m}^2/\text{s}^2)t}{\sqrt{(500.0 \text{ m}^2/\text{s}^2)t^2 + 100 \text{ m}^2}}$$; at $t = 10.0$ s this becomes 22.3 m/s.

2.73   The numbers meet after 100 s so the fly is in the air for 100 s so at 10 m/s it travels 1000 m, or to two figures 10×10^2 m.

2.83   (a) $v_H = (300 \text{ m/s})\cos 32.0° = 254$ m/s; (b) $v_V = (300 \text{ m/s})\sin 32.0° = 159$ m/s;

(c) $v = \sqrt{v_H^2 + v_V^2} = \sqrt{89\ 797}$ m/s $= 300$ m/s.

2.87   In 5.0 s the mouse moves 10 m north, hence the hypotenuse of the triangle is $\sqrt{(50 \text{ m})^2 + (10 \text{ m})^2} = 50.99$ m. $v = (51 \text{ m})/(5.0 \text{ s}) = 10.2$ m/s or 10 m/s. The dive angle is θ where $\tan\theta = (10 \text{ m})/(50 \text{ m})$ and $\theta = 11°$.

2.91   The two motions are perpendicular and $\mathbf{v}_{JE} = \mathbf{v}_{JS} + \mathbf{v}_{SE}$; $v_{JE} = \sqrt{v_{JS}^2 + v_{SE}^2} =$ $\sqrt{(10 \text{ km/h})^2 + (20 \text{ km/h})^2} = 22$ km/h. The angle the jogger makes measured from the direction of the ship is $\sin\theta = (10 \text{ km/h})/(22 \text{ km/h})$, $\theta = 27°$ which puts it 3° west of north.

Answers to Discussion Questions

3.3   The balls are identical and separately must fall together, that is, at the same rate. When two are stuck together and dropped alongside a single ball, they again must fall together provided the air resistance on the double mass is no different. In vacuum, there is no air resistance so both chunks of clay (one twice the mass of the other) must fall together. Indeed, 1000 such balls of clay stuck together and shaped into a piano would fall the same way — all things must fall together in vacuum.

3.7   The cars are always separated by one second. The distance each travels is proportional to the time squared so their separation will increase: after 1 s, the separation will be $\frac{1}{2}a(1\ s^2)$; after 10 s, it will be $\frac{1}{2}a[(10\ s)^2 - (9\ s)^2] = \frac{1}{2}a(19\ s^2)$.

3.9   If the acceleration were constant $v_{av} = \frac{1}{2}(v + v_0)$, and the average is midway between the initial and final speeds. But with a increasing, the average speed is closer to the initial speed than to the final speed.

3.11   The keys fall at the same rate as the floor and so cannot get any closer. The keys float in front of your face, which is also falling at the same rate.

3.15 $a = \left[(At^2 - Bt)/(t + C)D\right] + Et^2/(t - C)D$ when $t = 0$, $a = 0/CD + 0/$ $(-CD) = 0$; (b) $a = \left[(At^2 - Bt)/(tD)\right] + Et^2/(tD) = (At^2 - Bt + Et^2)/tD$; (c) $a = \left[(At^2 - Bt)/CD\right] - Et^2/CD = (At^2 - Bt - Et^2)/CD$.

3.17   The speed reaches a maximum of 4.2 s and then drops to zero at the peak altitude ($t_p \approx 7.4$ s). The net acceleration increases for about 2 s, after which it decreases as the engine throttles down. At $t = 4.2$ s, with the engine still firing, the net acceleration is zero. Thereafter it is increasingly negative as the engine's thrust decreases and gravity dominates. The engine shuts off at around 5.5 s and the rocket decelerates moving upward until it stops at $t \approx 7.4$ s.

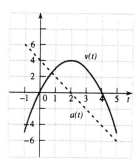

Answers to Multiple Choice Questions

1. a   3. c   5. b   7. b   9. e   11. a   13. d   15. c   17. d

Solutions to Problems

3.1 $\quad v = 100$ m/s, $v_0 = 0$; $a_{av} = \Delta v/\Delta t = (100$ m/s$)/(10$ s$) = 10$ m/s^2.

3.5 $\quad a_{av} = (v - v_0)/\Delta t = (15.0$ m/s $- 1.0$ m/s$)/(62$ s$) = 0.23$ m/s^2.

3.9 $\quad$ From Fig.3.2a, $a_{av} = $ constant $= a = \Delta v/\Delta t = (25$ m/s$)/(5.0$s$) = 5.0$ m/s^2.

3.33 $\quad v(t) = 3(12)t^2 - 2(6)t + 2.4$; $a_T(t) = 72t - 12$; which is zero at $t = (1/6)$ s.

3.51 $\quad v_0 = 2.2$ m/s, $v = 0$, $s = 10$ m; find a and then s during the 3rd second; $v^2 = v_0^2 + 2as$, $a = -v_0^2/2s = -(2.2$ m/s$)^2/2(10$ m$) = -0.242$ m/s^2. $s_2 = v_0 t + \frac{1}{2}at^2 = (2.2$ m/s$)(2.0$ s$) + \frac{1}{2}(-0.242$ m/s$^2)(2.0$ s$)^2 = 3.92$ m. $s_3 = (2.2$ m/s$)(3.0$ s$) + \frac{1}{2}(-0.242$ m/s$^2)(3.0$ s$)^2 = 5.51$ m. $s_3 - s_2 = 1.6$ m.

3.55 $\quad$ (a) In 1 s the first missile travels $s_1 = (60$ m/s$)(1$ s$) + \frac{1}{2}(20$ m/s$^2)(1$ s$)^2 = 70$ m. (b) $s_2 = 70$ m when $s_1 = (60$ m/s$)(2$ s$) + \frac{1}{2}(20$ m/s$^2)(2$ s$)^2 = 160$ m, hence $s_1 - s_2 = 160$ m $- 70$ m $= 90$ m.

3.67 $\quad v^2 = v_0^2 + 2gs = 0 + 2(9.81$ m/s$^2)(0.50$ m$) = 9.81$ m^2/s^2, $v_f = 3.1$ m/s $= 10$ ft/s $= 6.9$ mi/h.

3.85 $\quad v(t) = t^2 - t - 6$; $s(t) = \int_1^4 (t^2 - t - 6)\, dt = \left[\frac{1}{3}t^3 - \frac{t^2}{2} - 6t\right]_1^4 = (-10.7 + 6.17) = -4.5$ m

3.95

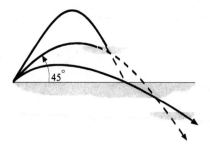

3.107 $\quad v(t) = t^2 - t - 6 = (t - 3)(t + 2)$ and therefore $v(t) \le 0$ when t is in the interval from 1 s to 3 s and $v(t) \ge 0$ when t is in the interval from 3 s to 4 s. Hence

$$l = \int_{t_1}^{t_2} |v(t)|\, dt = \int_1^3 -v(t)\, dt + \int_3^4 v(t)\, dt = 10.2 \text{ m}$$

Selected Answers to Discussion Questions

4.1   His ship was already moving rapidly so he should have shut off the engines and coasted, conserving fuel for the landing when he must change velocity. The writers didn't know the Law of Inertia.

4.5   The pellets rise, all the while slowing down due to drag and gravity. They come to rest (at a height of around 110 m) and then fall back, accelerating at a decreasing rate until they reach their terminal speed. Since the terminal speed is only about 9 m/s, the shot falls fairly harmlessly. By comparison, a .22-caliber bullet, because of its large mass, will drop much more rapidly and thus can be quite dangerous. Ordinary military bullets have still greater terminal speeds and are quite lethal when they return to ground level.

4.7   The idea was to fire the shell at a high angle (52°) so that it ascended into the stratosphere where the atmosphere was rarefied and the drag relatively weak. Actually 2 out of the 3.5 minutes of the flight were spent sailing through the thin air of the stratosphere.

4.11   When you catch the object, ordinarily your hand is free to move and bring it to rest in a relatively long time, which means that the force will be small for the same impulse and momentum change. Not so if your hand is against a table.

4.13   The more massive ball requires the greater impulse to bring it up to speed since doing so means a greater momentum change. Hitting your finger with a hammer is a matter of impulse and momentum-change and is independent of gravity. Yes, it would hurt.

Answers to Multiple Choice Questions

1. a   3. a   5. c   7. c   9. e   11. c   13. b   15. c   17. a
19. c

Solutions to Problems

4.3   The time of flight is t where $s_V = \frac{1}{2}gt^2$ and down is positive; $1.000\,0$ m =
$\frac{1}{2}(9.800\,0\ \text{m/s}^2)t^2$; $t = 0.451\,75$ s; and $s_H = v_H t = (2.213\,6\ \text{m/s})(0.451\,75\ \text{s}) = 1.000\,0$ m.

4.7   From Eq. (3.16) $s_R = -(2v_0^2/g)\cos\theta\sin\theta = \left[-(2)(405\ \text{m/s})^2/(-9.81\ \text{m/s}^2)\right] \times$
$\cos 45.0° \sin 45.0° = 16.7$ km.

4.19   If the tree is a distance s_H from the gun, the dart will arrive at $t = s_H/v_0\cos\theta$. At that moment it will have essentially dropped from the line-of-sight by a vertical distance of $\frac{1}{2}gt^2$, but that's the same drop the monkey experienced.

4.23   One force due east; the other two north and south of it at equal angles of $\pm\,\theta$ such that their N-S components cancel: 2 kN + (2 kN) cos θ + (2 kN) cos θ = 4 kN; 2 cos θ = 1, cos θ = 1/2, θ = $\pm 60°$

4.31   Unchanged, 10 kg·m/s.

4.39   The initial momentum of the cosmonaut-ship system is zero, $\mathbf{p}_i = \mathbf{p}_f = 0$ and $m_c v_{cf} + m_s v_{sf} = 0$, $v_s = -(60\ \text{kg})(10\ \text{m/s})/(5000\ \text{kg}) = -0.12$ m/s, where the direction of motion of the cosmonaut was positive.

4.45 $\Delta p = \displaystyle\int_{0\ \text{s}}^{6.00\ \text{s}} (15.0)t\ dt = \frac{1}{2}15.0(6.00)^2 = 270\ \text{N·s}; \Delta v = \Delta p/m = 270$ m/s.

4.55 $F_{av} = mgs_h/s_c$, since the landing is with both feet, the tolerable force is 100 kN; (60 kg)(9.81 m/s^2)s_h/(0.010 m) = 100 000; $s_h = 1.7$ m.

4.65 $p_i = p_f = 0$, (55.0 kg)v_f = (0.200 kg)(5.556 m/s); $v_f = 0.020\ 2$ m/s, in the opposite direction to the snowball.

Answers to Discussion Questions

5.1   The bullet will have momentum and can punch a hole in a silver dollar that has mass and therefore inertia. The silk has so little mass and inertia that the bullet will push it out of the way rather than go through it.

5.9   The more massive the leg, the more force required to accelerate and decelerate it as the animal runs. Dogs and deer have thin lower legs, each supporting one-quarter of the animal's weight.

5.11   The water cuts down considerably on the coefficient of friction (see Problem 5.61). By comparison, it would be much harder to skate on dry glass. Skis are waxed to reduce the coefficient of friction. Wet ice is slippery. The marbles would allow the coffin to be rolled into place rather than slid, and the coefficient of rolling friction is smaller than the coefficient of kinetic friction.

5.13   The locked wheels slide and so have less friction. Any slight asymmetry and the car will spin around so the locked wheels, which are less inhibited, lead the way down the incline. If the brakes cannot be made to engage all at once, it would be best to have the front wheels engage first, to reduce the possibility of the car sluing around 180°.

Answers to Multiple Choice Questions

   1. c   3. d   5. e   7. c   9. b   11. c   13. d   15. d   17. a
19. c

Solutions to Problems

5.3 $F_W = mg = (0.50 \text{ kg})(9.8 \text{ m/s}^2) = 4.9$ N, anything in excess of this.

5.13 $F(t) = ma(t) = m\dfrac{dv}{dt} = m(6t - 12)$.

5.21 $\Sigma F_H = F_T \sin\theta = ma$, $\Sigma F_V = F_T \cos\theta - F_W = 0$ and $F_T \cos\theta = F_W$
dividing these two equations, $\tan\theta = a/g$; $a = (26.8 \text{ m/s})/(6.8 \text{ s}) = 3.94 \text{ m/s}^2$; $\tan\theta = 0.402$; $\theta = 22^\circ$.

5.25 $\Sigma F_{\parallel} = F_W \sin\theta = ma$, $mg \sin\theta = ma$, $a = 0.342g$. $v^2 = 2as = 2(0.342)(9.8 \text{ m/s}^2) \times$
(20 m), $v = 12$ m/s.

5.33 $F = ma$; $100.0a = 120 \times 10^2 t + 40.0 \times 10^2$; $dv/dt = a = 1.20 \times 10^2 t + 0.40 \times 10^2$;

$$\int_{2.00}^{v} dv = \int_{0}^{t} 1.20 \times 10^2 t \; dt + \int_{0}^{t} 0.040 \times 10^2 \, dt; \; v(t) = (60.0 \text{ m/s}^3)t^2 + (40.0 \text{ m/s}^2)t + 2.00 \text{ m/s}.$$

5.37 $m_1 = m_2 = m$; $F_{W1} - F_T = m_1 a$, $(m + m_g) g_M - F_T = (m + m_g)a$; $F_T - F_{W2} = ma$ (where $F_{W2} = mg_M$). Adding the last two equations; $(m + m_g) g_M - mg_M = (2m + m_g)a$ and so $g_M = (2m + m_g)a/m_g$. Now to find a. $v = (1.2 \text{ m})/(3.0 \text{ s}) = 0.40$ m/s; $v^2 = 2as$; $a = v^2/2s = (0.40 \text{ m/s})^2/2(0.50 \text{ m})$, $a = 0.16 \text{ m/s}^2$, $g_M = (0.50 \text{ kg} + 0.025 \text{ kg})(0.16 \text{ m/s}^2)/ (0.025 \text{ kg}) = 3.4 \text{ m/s}^2$.

5.49 $v^2/R = a_c = 9.0g$, $(290 \text{ m/s})^2/\left[(9.0)(9.8 \text{ m/s}^2)\right] = R$, $R = 9.5 \times 10^2$ m.

5.57 $F_N = m(v^2/r - g)$; here $F_N = F_W/2 = mg/2 = m(v^2/r - g)$, $v = \sqrt{3gr/2} = 19.2$ m/s.

5.67   From Eq. (5.14) $\tan \theta_{max} = \mu_S$, $\tan 17° = 0.31$.

5.75   Taking the direction of motion as positive, $\Sigma F_H = -F_f = ma$, $\mu_S F_W = -ma$, $a = -\mu_S g$; $v = v_0 + at$, $v_0 = -at$, $t = +v_0/\mu_S g = (27 \text{ m/s})/0.9(9.8 \text{ m/s}^2) = 3$ s.

5.83 $\Sigma F_{\parallel} = F_f - F_W \sin \theta = 0$; $\mu_S F_N = F_W \sin \theta$, $\mu_S(0.43 \; F_W)\cos \theta = F_W \sin \theta$, $0.43 \; \mu_S = \tan \theta$, $\tan \theta = 0.387$, $\theta = 21° = 0.2 \times 10^2$ degrees.

Answers to Discussion Questions

6.1   (a) If the three forces are equal and add up to zero, they must be 120° apart, coplanar and concurrent. (b) Another way to see this concept is to realize that the three vectors, which supposedly add up to zero, must form a closed triangle when placed tip-to-tail. These three cannot be the sides of any such triangle since (30 N + 40 N) can at most add vectorially to produce a vector of magnitude 70 N, not 80 N.

6.5   The arm pivots on the end of the humerus. To raise a load, the biceps contracts, while the triceps stays relaxed; to push down, the triceps contracts, while the biceps relaxes.

6.9   (a) Ordinarily, when touching your toes, you shift the lower portion of your body back as the upper portion leans forward, keeping the *c.g.* over your feet. (b) To stand on your toes, you must move your *c.g.* over your toes so it will be supported. Here, the edge of the door stops you from shifting your mass.

6.15   Held tightly together, the boards bend concavely around the fist. The front face of the front board is in compression, while the back board is in tension, being stretched into a curve. The wood along the grain is weak in tension, and the rear board develops a crack at the back face and shatters in two.

6.17   Your arms will be in tension, your legs in compression.

Answers to Multiple Choice Questions

   1. d   3. d   5. c   7. e   9. b   11. a   13. e   15. d   17. d
   19. c

Solutions to Problems

6.1   (a) 250 N (b) 200 N.

6.11 $+\uparrow \Sigma F_V = (15.0\ \text{N}) \cos 45.0° + (31.0\ \text{N}) \cos 20.0° - F_{W3} = 0$; and
$F_{W3} = 39.7\ \text{N}$. $(15.0\ \text{N}) \sin 45.0° = (31.0\ \text{N}) \sin 20.0°$.

6.17 $F_T = (1.0\ \text{kg})g$; $+\uparrow \Sigma F_V = 2F_T \sin 20° - F_W = 0$; $F_W = 6.7\ \text{N}$.

6.23 $F_{T1} = (80.0\ \text{kg})g = 784.5\ \text{N}$; $F_{T2} = (10.0\ \text{kg})g = 98.07\ \text{N}$; $\overset{+}{\rightarrow} \Sigma F_H =$
784.5 N $- F_f - 98.07$ N $= 0$; $F_f = 686\ \text{N} = \mu_s F_N$; $\mu_s = 0.93$, it makes no sense to give μ_s to any more than 2 significant figures.

6.27 $\tau_0 = (1.00 \text{ m})(100 \text{ N}) = 100 \text{ N} \cdot \text{m}.$

6.37   Taking the reaction forces as F_{RA} and at the scale, F_{RB} it follows that $(+\Sigma \tau_A =$
$g(1.0 \text{ kg})(0.40 \text{ m}) - F_{RB}(0.20 \text{ m}) - g(2.0 \text{ kg})(0.10 \text{ m}) = 0$ and $F_{RB}/g = 1.0 \text{ kg}.$

6.51 $\underset{\rightarrow}{+}\Sigma F_H = F_{T2} \cos 60.0^\circ - (100 \text{ N}) \cos 60.0^\circ = 0;\ F_{T2} = 100 \text{ N};\ +\uparrow\Sigma F_V =$
$2(100 \text{ N}) \sin 60.0^\circ - F_W = 0;\ F_W = 173 \text{ N}.$ The line-of-action of each of the two tension
forces intersect at a point directly below the *c.g.* since F_W must pass through that point too.

6.61   About the toes, $(+\Sigma \tau = 0 = 2 F_{Nh}(1.50 \text{ m}) - g(65 \text{ kg})(1.00 \text{ m});\ F_{Nh} = 0.21 \text{ kN}$ on
each hand. $+\uparrow\Sigma F_V = 2(212 \text{ N}) + 2F_{Nt} - g(65 \text{ kg}) = 0;\ F_{Nt} = 0.11 \text{ kN}$, on each foot.

Answers to Discussion Questions

7.3   The gravitational field inside a uniform spherical shell of mass is zero. (See Fig.7.5.) That would not be the case if the mass were not completely symmetrical. There is no shielding of the inside, and new arrivals would be detected gravitationally.

7.5   When the craft is tilted, the little mass at the end of the boom and the main portion of the satellite both tend to rotate about the craft's *c.g.* as they fall into the gravitational field. The Earth's gravity drops off as $1/r^2$ and so, if the small mass at the bottom of the boom (which experiences a greater field), develops a greater torque, it will drive the boom back to being vertical. The craft will thereby remain stable pointing downward all the time.

7.7   The Earth rotates counterclockwise looking down onto the North Pole, which means that any rocket on the launch pad is already moving eastward at the speed of the Earth even before lift-off. The closer to the equator, the greater the speed of a point on the Earth's surface. So, launch from a southerly location and launch eastward and you get as much benefit from the planet's rotation as possible.

7.11   The rocket fires with a forward thrust and that carries it into an elongated elliptical orbit. The additional speed allows it to move outward beyond the circular orbit. It slows down as it moves away from the center of force. As it reaches its most distant point (just before it would begin to fall back inward), the rocket fires again, giving it enough speed to satisfy Eq. (7.7) and go into a large circular orbit. Despite all maneuvering, the final orbital speed is less than the initial orbital speed.

7.15   These antennas pick up TV signals from satellites parked in geosynchronous orbits. In the Northern Hemisphere, these are all seen in the southern part of the sky since they are over the equator. Each craft is fixed at a known location in space so you need only decide which one you would like to tune in.

7.17   Since gravity increases as the mass of a body increases, there will come a point where the gravitational force tending to squash it exceeds all structural forces keeping it from deforming. Beyond that limit, which is around 1000 km in diameter, a typical solid celestial body will become essentially spherical. Apparently, Hyperion is just too small.

Answers to Multiple Choice Questions

  1. c   3. d   5. c   7. c   9. b   11. d   13. c   15. b

Solutions to Problems

7.1 $\quad F'_W = G(2m)M_\oplus/(2R_\oplus)^2 = F_W/2.$

7.7 $\quad F_G = GM_V M_\oplus/r^2_{V\oplus} = 2.5 \times 10^{16}$ N.

7.19 $\quad mg_\oplus = GmM_\oplus/r^2$ and $g_0 = GM_\oplus/R^2_\oplus$ hence $g_\oplus = g_0 R^2_\oplus/r^2.$

7.33 $\quad GM_n/R^2_n = v^2/R_n,\ [G\,(4/3)\pi R^3_n \rho]/R^2_n = (2\pi R_n/T)^2/R_n,\ (G\rho/3\pi)^{1/2} = 1/T,$
$T = 1 \times 10^{-3}$ s.

7.41 $\quad$ From Eq. (7.8), $v_0 = \sqrt{GM_c/r_c} = [(6.67 \times 10^{-11}\ \text{N·m}^2\text{·kg}^{-2})(7.35 \times 10^{22}\ \text{kg})/(18.0 \times 10^5\ \text{m})]^{1/2} = 1.65 \times 10^3$ m/s.

7.49 $\quad \mathfrak{g}_\odot = GM_\odot/r^2_\odot = G(2.0 \times 10^{30}\ \text{kg})/(1.5 \times 10^{11}\ \text{m})^2 = 5.9 \times 10^{-3}$ m/s$^2.$

7.57 $\quad F_G = GmMr/R^3;\ \mathfrak{g} = GMr/R^3,\ r = R/2$ hence $\mathfrak{g} = GM/2R^2.$

Answers to Discussion Questions

8.1   The outside wheels turn faster. A differential gear is mounted between pairs of wheels.

8.5   Yes. If the moment-arm is large, the torque will be large. No. We cannot conclude anything about α and $\Sigma \tau$.

8.7   It must be zero since $\mathbf{L}$ = constant, $\Delta\mathbf{L}$ = 0 and τ = 0.

8.9   The fact that the line-of-action of the force does not pass through the *c.m.* results in a torque; $\mathbf{L}$, ω, and τ are up out of the page. There will be a slight rise of the *c.m.* and then a ballistic flight. She gains angular momentum and will flip over.

8.13   *L* in the ball-person system is not conserved because there's a friction force via the ground on the feet. Enlarging the system to include the Earth conserves angular momentum, and so the Earth's motion must change accordingly. The person would both rotate and translate.

8.15   At a constant speed, the forward propeller reaction force $\mathbf{F}$, which is beneath the *c.m.*, equals the friction force arising from the air and water, just as the weight equals the upward buoyant force, and so $\Sigma\mathbf{F}$ = 0. At any moment (neglecting friction and any vertical component from the propeller force), there is a balance between the torque from the propeller force and the buoyant force. The greater $\mathbf{F}$ is made, the more the boat noses up and the farther toward the rear (away from the *c.m.*) the point of action of the buoyant force moves, thereby increasing its torque until equilibrium is reached again.

8.19   Nothing.

Answers to Multiple Choice Questions

   1. b   3. e   5. b   7. c   9. b   11. b   13. d   15. a   17. d
   19. b

Solutions to Problems

8.7   The distance between adjacent lines $\ell = (0.30 \text{ m})/525 = 0.571 \times 10^{-3}$ m. $\ell = r\theta$; $r = (0.571 \times 10^{-3} \text{ m})/\theta$; $\theta = 0.01667° = 0.00029$ rad; $r = 1.96$ m or 2.0 m.

8.15 $\omega = \alpha t$; $(500 \text{ s}^{-1})(2\pi)/(30 \text{ s}) = \alpha$, $\alpha = 105$ rad/s^2.

8.29 $\omega = v/R = (1.467 \text{ ft/s per mi/h})(20 \text{ mi/h})(1000 \text{ ft}) = 0.029$ rad/s. $a_c = R\omega^2 = (1000 \text{ ft})(0.029 \text{ rad/s})^2 = 0.86$ ft/s$^2 = 0.26$ m/s^2.

8.33   The speed of a point on the belt is $R_A \theta_A = \ell_B = R_C \theta_C$, $\ell_E = R_D \theta_C$; $0.3 \theta_A = 0.1 \theta_C$ and $\omega_C = (0.3/0.1)\omega_A = 3\omega_A$, $\ell_E = 0.4 \theta_C$ and so $v_E = 0.4 \ \omega_C = 1.2 \ \omega_A$; $v_E = 1.2 \times 2\pi/60 = 0.13$ m/s, and to one significant figure $v_E = 0.1$ m/s, upward.

8.47   33⅓ rpm = 3.49 rad/s, $v_B = R_p \omega_p = R_t \omega_t$ where it's p for pulley and t for turntable. $\omega_t = \alpha_t(6.0 \text{ s})$; $\omega_t = 3.49$ rad/s; $\alpha_t = 0.58$ rad/s^2; $\alpha_p = 15\alpha_t = 8.7$ rad/s^2. $\omega_p = 52.4$ rad/s. $\theta = \frac{1}{2}\alpha t^2 = \frac{1}{2} 8.7 \ (\text{rad/s}^2)(36 \ \text{s}^2) = 156.6$ rad = 25 rev.

8.49 $\omega_i = 2\pi/T$, $\omega_f = 2\pi/(T + \Delta T)$; $\Delta\omega = -2\pi/T + 2\pi/(T + \Delta T) = -2\pi\Delta T/T(T + \Delta T) \approx -2\pi\Delta T/T^2$. $\alpha = \Delta\omega/\Delta t = \Delta\omega/T = -2\pi\Delta T/T^3 = -2\pi(25 \times 10^{-9} \text{ s})/(24 \text{ h} \times 60 \times 60)^3 = -2.4 \times 10^{-22}$ rad/s^2. $\omega_f = \omega_i + \alpha t$, $\Delta\omega = -7.6 \times 10^{-6}$ rad/s and $\omega_f = \omega_i + \Delta\omega = (2\pi/24 \times 60 \times 60) - 7.6 \times 10^{-6}$ rad/s = 6.51×10^{-5} rad/s. $T_f = 2\pi/\omega_f = 9.65 \times 10^4$ s = 27 h.

8.69 $+\downarrow\Sigma F_V = ma = F_W - F_T$; $\circlearrowleft \Sigma\tau = I\alpha = F_T R$; $a = R\alpha$; $F_T = I\alpha/R = Ia/R^2$, $F_W - Ia/R^2 = ma$, $a(m + I/R^2) = mg$, $a = mg/(m + I/R^2)$.

8.77 $\mu_s = F_f/F_N = (2/7)mg \sin\theta/mg \cos\theta = (2/7)\tan\theta$.

8.85 $r_1 = l/200$, $r_2 = l/100 + l/200$, $r_3 = 2l/100 + l/200$, ... , $r_n = 99l/100 + l/200$; $I = \Sigma mr^2 = (M/100)[(l/200)^2 + (3l/200)^2 + (5l/200)^2 + ... + (199l/200)^2] = (Ml^2/4 \times 10^6)(1^2 + 3^2 + 5^2 + ...+ 199^2]$; $[1^2 + 3^2 + 5^2 + ...+ 199^2) = (1/3)100 \ (201)(199)$; hence, $I \approx (1/3)Ml^2$.

Answers to Discussion Questions

9.1   No work is done if the rocket doesn't move. The rocket pushes down on the exhaust gas and the gas pushes up on the rocket, accelerating it. Momentum is conserved, so the momentum of the vehicle equals the momentum of the gas. The work done on the rocket increases with time as it moves faster and therefore farther per second. With respect to the ground, the gas initially gets most of the KE. As the vehicle speeds up, more energy is transferred to it. All the KE, rocket and exhaust, comes from the chemical energy of the fuel.

9.3   (a) He is overcoming air drag. (b) The reaction of the ground on him is the propelling force. If his feet slide back, the forward-directed force of the-floor-on-him will do negative work on him. He will do work heating the floor. (c) If his feet do not slip, he does no work on the floor and the floor does no work on him. (d) His KE comes from his food.

9.5   (a) He did work on the bike and provided the energy that got things started. (b) The reaction force of the ground. (c) No, with no slipping there's no relative motion between the tire and ground, and no work is done on the bike by the road. Even when there is distortion of the tire and road, the work is done on the road. The KE of bike and rider comes from the internal energy of the rider.

9.7   Assuming both balls are launched with the same very slow speed, the ball moving along the depression wins the race. Its velocity will increase as it descends and therefore its component in the forward direction, parallel to the plane, will always be equal to or greater than the other ball's. This will continue to be the case up to some threshold launch speed. See *Phys. Teach.* 33, 376 (1995).

9.9   Work is energy transferred to or from a system via the application of a force acting over a distance. Historically, it got its name before the concept of energy was formalized, and so it continued to be treated as if it were something other than energy. Ergo, the notion that energy is the ability to do work. It's just as circular as saying energy is the ability to produce *vis viva* (KE). It makes sense, since work and PE both have the ability to produce *vis viva*, but it's clearly circular.

9.11   At a constant speed there is a force, but it's perpendicular to *s*. When the ball has a tangential acceleration there is a tangential force — work done — and an increase (or decrease) in KE. A tangential component of the tension can exist when the string leads the ball, making an angle with **v** of less than 90°.

9.13   (a) At the top of the first hill. (b) You want the lowest, steepest drop right at the beginning of the ride. (c) No. (d) Yes, provided the maximum KE does not occur when PE = 0; make each valley lower than the one before by just enough to regain the KE lost to friction. (e) No. (f) They all have to be lower than the first.

9.17   The friction force accelerates the crate such that $W = \Delta KE = \frac{1}{2}mv^2$. The work is positive. The work overcomes inertia. The energy appears as KE. When stopping, the friction force is in the opposite direction to the displacement and does negative work, converting KE to thermal energy.

Answers to Multiple Choice Questions

1. a   3. a   5. a   7. c   9. c   11. a   13. a   15. c   17. e
19. c   21. d

Solutions to Problems

9.13   There are 5 ropes supporting the load hence 5.0 m of rope will have to be pulled out.

9.21 $W = \mu_r mg (\cos \theta) l = (4.9 \text{ N})(\cos 10^\circ)(25 \text{ m}) = 1.2 \times 10^2 \text{ J}$.

9.29 $(0.40 \text{ liter/min})(2.0 \times 10^4 \text{ J/liter}) = 8.0 \times 10^3 \text{ J/min} = 1.3 \times 10^2 \text{ W}; BMR = (1.3 \times 10^3 \text{ W})/(1.8 \text{ m}^2) = 74 \text{ W/m}^2$.

9.43   From Conservation of Momentum $v_G = m_b v_b / m_G = 0.98$ m/s and so $KE_G = 0.95$ J while $KE_b = 0.29$ kJ.

9.49 $$W = \int F \, dl = \int m \frac{dv}{dt} \, dl = m \int v \frac{dv}{dl} \, dl = m \int_{v_i}^{v_f} v \, dv$$

9.59   The 100-N weight doesn't move. (a) $W = (10 \text{ N})(10 \text{ m}) = 100$ J. (b) the weight is too heavy to rise so no rope is on the floor. (c) $\Delta PE = +100$ J.

9.69   The work done by gravity is the change in the KE, $W_G = \Delta KE$; the component of the weight acting down the incline is opposite to l; $-(F_w \sin \theta) l = \frac{1}{2}(mv_f^2 + I\omega_f^2) - \frac{1}{2}(mv_i^2 + I\omega_i^2)$ but $v_f = \omega_f = 0$ and $v = R\omega$; $I = (2/5)mR^2 = 8.0 \times 10^{-3} \text{ kg} \cdot \text{m}^2$; $19.6 \text{ N} \times 0.342 \times l = \frac{1}{2}(7/5)mv_i^2 = 35$ J; $l = 5.2$ m.

9.73 $\Delta PE = mg\Delta h = (55.0 \text{ kg})(9.80 \text{ m/s}^2)(9.00 \text{ m}) = 4.85 \times 10^3$ J. This is the energy stored in the trampoline. $h = 9.00$ m or back up to 10.0 m above ground.

9.81   Compare the two values of kinetic energy at escape speeds; $v_{esc}^2(\text{Moon})/v_{esc}^2(\text{Earth}) = (2.4 \text{ km/s})^2/(11.2 \text{ km/s})^2 = 0.046 \approx 5\%$.

9.93   The velocity vector v_i of the moving ball has two components, $v_{i\parallel}$ along the line-of-centers and $v_{i\perp}$ perpendicular to that. The result is as if the two motions occurred independently — the perpendicular motion constitutes a miss while the parallel motion transfers all that momentum to the target ball, which sails off along the line-of-centers while the incident ball, no longer possessing momentum in that direction, moves away perpendicular to the line-of-centers.

Answers to Discussion Questions

10.1   Aluminum — foil, pots; tungsten — light bulbs, in steel knife blades; carbon — in wood, pencil lead, diamond; mercury — in fluorescent bulbs, silent switches, thermometers; zinc — paint pigment, ointments, in brass; chlorine — bleach, drinking water, salt; copper — pots, wire, jewelry, in brass and bronze; sodium — table salt, Alka-Seltzer; iron — blood, raisins, nails, pots; lead — plumbing, solder, old paint; magnesium — griddles, antacid pills, sparklers; cobalt — blue dyes and pigments; oxygen — air, water, rust; chromium — in steel, plated on toasters and car bumpers; nickel — alloyed in coins, in common Alnico magnets.

10.3   (a) low-carbon steel (b) tempered steel (c) all have the same slope, same Y (d) tempered steel (e) tempered steel (f) low-carbon steel (g) all are the same (h) low-carbon steel.

10.5   (a) The area under the force-displacement curve is the work done on the sample, the energy mechanically entered into the material in the process of distorting it. (b) Since stress is force over area, and strain is displacement over length, the product of the two is the strain energy divided by the volume. It has the units of J/m^3. (c) The colored area corresponds to the increased internal energy per unit volume of the sample.

10.7   Both ropes break when the stress exceeds a certain value that is the same for each, so they can carry the same load. But more work will have to be done to break the longer rope (which elongates more) than the shorter one.

10.9   The stress in the broad base, which carries the entire load, is less, which is one reason why ancient walls tapered upward.

10.11   Elephants have to walk around very carefully (they don't do much jumping). The compressive strength (Table 1) of a horse femur is only 145 MPa. We have bred horses to the point where they are almost too big for their bones — at least with all the jumping and running we demand of them.

10.13   More work goes into the rubber than comes out, and the difference (the area of the colored region in the figure) appears mostly as thermal energy. Rubber is therefore useful to convert unwanted mechanical energy into thermal energy. When a rubber band is worked cyclically, it will get noticeably warmer.

10.15   Human tendon has to be able to store a great deal of energy without permanently distorting — it's highly resilient, twice as much as spring steel. Bridges and springs must be resilient; they must recover after being compressed or stretched. Glass has very little ductility. The area under its stress-strain curve is small and it has little toughness — that's why a glass or china plate will shatter when dropped. A bow or pole should be resilient.

Answers to Multiple Choice Questions

1. d   3. d   5. b   7. d   9. c   11. b   13. b   15. a   17. c
19. b

Solutions to Problems

10.3 $2000 \text{ lb/in.}^3 = (907.2 \text{ kg})/(2.54 \times 10^{-2} \text{ m/in.})^3 = (907.2 \text{ kg})/(16.39 \times 10^{-6} \text{ m}^3) = 6 \times 10^7 \text{ kg/m}^3$.

10.17 $N_A \, \rho/M_m = (6.022 \times 10^{26})(1.000 \times 10^3 \text{ kg/m}^3)/(18 \text{ kg}) = 3.3 \times 10^{28}$ molecules/m^3.

10.21   Its molecular mass is $2(12 \text{ u}) + 5(1 \text{ u}) + 16 \text{ u} + 1 \text{ u} = 46 \text{ u}$ hence from Eq. (10.2), $L = [(46 \text{ g/mol})/(6.022 \times 10^{23} \text{ molecules/mole})(0.789 \text{ g/cm}^3)]^{1/3} = 4.59 \times 10^{-8} \text{ cm} = 0.46 \text{ nm}$.

10.31 $\epsilon = \Delta L/L_0 = (1.5 \times 10^{-2} \text{ m})/(10 \text{ m}) = 0.15\%$.

10.39 $P = F/A = (10 \text{ N})/(0.20 \text{ m} \times 0.30 \text{ m}) = 1.7 \times 10^2 \text{ N/m}^2$.

10.45 $\Delta PE_e = \frac{1}{2}ks^2$; $k = 2(3.2 \times 10^{-19} \text{ J})/(0.2 \times 10^{-9} \text{ m})^2 = 16 \text{ J/m}^2$.

10.49 $A = \pi R^2 = \pi(1.25 \times 10^{-2} \text{ m})^2 = 4.91 \times 10^{-4} \text{ m}^2$; $\sigma_R = F/A = 15 \text{ MPa}$. The new area is four times the old hence $\sigma_R 4A = 4(7363 \text{ N}) = 29.5 \text{ kN}$.

10.53 $F = 25\,000 \text{ lb} = 111 \text{ kN} = A\sigma = \pi R^2 \sigma$; $(111 \text{ kN})/\pi(345 \text{ MPa}) = R^2$; in excess of $R = 1.01 \text{ cm}$. $\epsilon = \sigma/Y = (345 \text{ MPa})(200 \text{ GPa}) = 0.173\%$.

10.55   The total area being sheared is $5A = 5\pi R^2 = 6.28 \times 10^{-3} \text{ m}^2$; $\sigma = F/A$; $F = \sigma A = (145 \text{ MPa})(6.28 \times 10^{-3} \text{ m}^2) = 0.911 \text{ MN}$.

Answers to Discussion Questions

11.1   (a) Increase as the effective g increases. (b) Zero.

11.5   Work is done filling the balloon and displacing the atmosphere. Gravitational-PE is stored in the balloon-atmosphere system. As the balloon ascends, the atmosphere descends.

11.7   (a) Nothing — you now displace one-cup's weight more of water, so the level stays the same. (b) The bust floats partially submerged, displacing its weight of water. When it was aboard, it was displacing its weight as well, so the level again does not change. (c) Nothing. (d) The level drops.

11.11 $P_1 = P_T + \rho gh$ hence $P_T = P_1 - \rho gh$ and when $P_1 = 0$, $P_T < 0$.

11.13   The soap drops the surface tension in the middle and the surface pulls out to the periphery, as if you had punched a hole in a stretched rubber sheet.

11.15   Air streaming over it asymmetrically (the ball drops a little, allowing most of the air to rush over it) produces a pressure drop above the ball (where a lot of air is moving rapidly), resulting in lift.

11.17   Air beneath the ball will move more rapidly than above; a pressure drop below the ball will cause it to sink more swiftly than if it was not spinning.

11.19   To have lift, the wing must have air circulating around it, which means angular momentum. Somehow, an equal and opposite amount of angular momentum will be imparted to the air-plane system and that's done nicely with the creation of a starting vortex revolving in the opposite direction. If the wing is stopped, the circulating air will generate another vortex equal and opposite to the starting vortex.

Answers to Multiple Choice Questions

1. b   3. d   5. a   7. a   9. a   11. d   13. b   15. c   17. c
19. d

Solutions to Problems

11.5 $1.000 \text{ lb/in}^2 = (4.44823 \text{ N})/(1.000 \text{ in}^2)(2.540 \times 10^{-2} \text{ m/in.})^2 = 6.895 \times 10^3 \text{ N/m}^2$.

11.13 $P_iV_i = P_fV_f$; $(1.013 \times 10^5 \text{ Pa})(10 \text{ cm}^3) = P_f(2.5 \text{ cm}^3)$; $P_f = 4.1 \times 10^5 \text{ Pa}$.

11.19 $P = P_s + \rho gz$; $dP/dz = \rho g$.

11.27 $P_G = \rho gh = -(1.00 \times 10^3 \text{ kg/m}^3)(9.8 \text{ m/s}^2)(1.1 \text{ m}) = -1.1 \times 10^4 \text{ Pa}.$

11.33   The weight of the sea water displaced (of volume V_w) equals the weight of the entire berg (of volume V_i) hence $V_w(1.025 \times 10^3 \text{ kg/m}^3)g = V_i(0.92 \times 10^3 \text{ kg/m}^3)g$; $V_w/V_i = 89.8\%$ or 90% is submerged and 10% is visible. For fresh water $V_w/V_i = 92\%$ and 8% is above.

11.41   Volume of 2.54 cm section is $(3.05 \text{ m})(6.10 \text{ m}) \times (2.54 \times 10^{-2} \text{ m}) = 0.473 \text{ m}^3$ and that much water weighs $(0.473 \text{ m}^3)(1.00 \times 10^3 \text{ kg/m}^3)g = 4.63 \text{ kN}$ (per 2.54 cm of settle). The raft weighs $(5.67 \text{ m}^3)(0.50 \times 10^3 \text{ kg/m}^3)g = 27.8 \text{ kN}$ which equals a volume of water of $V_w(1.00 \times 10^3 \text{ kg/m}^3)g$ or $V_w = 2.84 \text{ m}^3$ and that's a depth of $(2.84 \text{ m}^3)/(3.05 \text{ m})(6.10 \text{ m}) = 0.15 \text{ m}.$

11.59 $\Delta P = \rho gh = (1.05 \times 10^3 \text{ kg/m}^3)(9.81 \text{ m/s}^2)(0.85 \text{ m}) = 8.75 \text{ kPa}$ or 8.8 kPa as compared to 13.3 kPa at the heart and so the reduction is significant.

11.63   To fall a height $y = \frac{1}{2}gt^2$ takes a time $t = \sqrt{2y/g}$ hence traveling horizontally, $x = \sqrt{2gh}\,\sqrt{2y/g} = 2\sqrt{yh}.$

11.67   From Bernoulli's Equation, $P_0 + \frac{1}{2}\rho v^2 = P$; $\Delta P = \frac{1}{2}\rho v^2$; $v = \sqrt{2\,\Delta P/\rho}.$

11.69   From Eq. (11.22), $P_p - P_t = \frac{1}{2}\rho v_t^2(A_p^2 - A_t^2)/A_p^2$. Since $\Delta P = \rho g\Delta y$ and from the Continuity Equation $A_p v_p = A_t v_t$, consequently $\rho g\Delta y = \frac{1}{2}\rho v_p^2(A_p^2 - A_t^2)/A_t^2$; $v_p = \sqrt{2g\Delta y/[(A_p^2/A_t^2) - 1]}.$

11.71   From Bernoulli's Equation; $P_i - P_A = \frac{1}{2}\rho v^2$. $(3.79 \text{ MPa} - 0.101 \text{ MPa})2/\rho = v^2$; $v = 82 \text{ m/s}$, or about 180 mph.

11.75 $P_1 + \frac{1}{2}\rho v_1^2 + \rho gy_1 = P_2 + \frac{1}{2}\rho v_2^2 + \rho gy_2$, using gauge pressure and taking $y = 0$ at the hole, $0 + 0 + \rho g(H + h) = \rho gh + \frac{1}{2}\rho v_2^2 + 0$; $v_2 = \sqrt{2gH}.$

Answers to Discussion Questions

12.1   When θ is a maximum $v = 0$, as is KE, a is a maximum as is PE. When $\theta = 0$, $a = 0$, as is PE, but v is maximum, as is KE.

12.3   The force tending to stretch the spring is a maximum twice during each cycle of the pendulum (viz., when it's vertical). Thus the frequency of the spring oscillation should be made equal to twice the pendulum frequency. See W.R. Mellen, *Phys. Teach. 32,* 122 (1994).

12.5   Increasing the mass will decrease the natural frequency.

12.7   The parachute will swing like a pendulum and as the vortex frequency matches the pendulum frequency, the resulting resonance will send the rider wildly swinging.

12.9   Like the bob on a helical spring, the greater the mass the smaller the frequency.

12.11   At any moment (when at a distance r from the center of Mongo), the weight of the ball depends on the mass M of material remaining "beneath" it. Only the mass of the sphere of matter of radius r attracts the ball. $F_G \propto mM/r^2$, where M is proportional to the volume, which goes as r^3. Hence, $F_G \propto r$ and the motion must be SHM.

12.13   The waves vanish, which certainly means there is no elastic-PE stored in the undistorted rope. All the energy is kinetic, though undisplaced at that very instant the various segments of the rope in the region of overlap are nonetheless moving vertically.

12.15   The speed of the wavepulse varies with the square root of the tension, which, in turn, is determined by the load and the weight of the string itself. The tension increases all the way up to the point of support because of the linear mass density of the string. Hence, the speed increases as the wave rises. Air friction and internal losses will convert some of the energy of the wave to thermal energy and its amplitude will diminish.

12.17   The materials are getting progressively more rigid as we go from clay down the table. Just compare the softness of clay to the hardness of granite. Accordingly, we can expect that the internal restoring force will increase as the interatomic force increases, and so the speed of a compression wave will increase. Of course, an increase in rigidity corresponds to an increase in Young's Modulus.

12.19

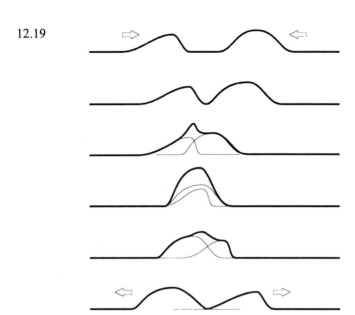

Answers to Multiple Choice Questions

1. b   3. a   5. c   7. c   9. d   11. c   13. a   15. d   17. d
19. b   21. c

Solutions to Problems

12.3   The motion is along a line and SHM. $f = $ (78 rot/min)/(60 s/min) = 1.3 Hz. $\omega = 2\pi f = 2.6\pi$ rad/s.

12.19 $F = mg = ky$; $k = 784$ N/m. $f = (1/2\pi)\sqrt{k/m} = 3.15$ Hz.

12.35   From Eq.(12.5) and (12.7) at $t = 0$, $x_i = A \cos \epsilon$, $v_i = -A\omega_0 \sin \epsilon$; now divide. If the initial speed is zero $\tan \epsilon = 0$ and $\epsilon = 0$ or whole number multiples of π. Now square both equations and add. At the initial position the speed is nonzero so KE $\neq 0$ so the PE is not a maximum and hence $x_i \neq A$.

12.39 $mg = k(0.020 \text{ m})$, $k/m = g/(0.020 \text{ m})$ and $f = (1/2\pi)\sqrt{g/(0.020 \text{ m})} = 3.5$ Hz.

12.43   Energy is not conserved though momentum is; $m_b v_b = (m_b + M)v_0$; $v_0 = 2\pi f x_0 = 8.48$ m/s; $(0.005\ 0 \text{ kg})v_b = (0.505 \text{ kg})(8.48 \text{ m/s})$ and so $v_b = 0.86$ km/s.

12.55 $dE/dt = 0$; $\dfrac{d}{dt}\left[\dfrac{1}{2}m\left(\dfrac{dx}{dt}\right)^2 + \dfrac{1}{2}kx^2\right] = 0$ hence $\left(\dfrac{1}{2}m\right)2\left(\dfrac{dx}{dt}\right)\dfrac{d}{dt}\left(\dfrac{dx}{dt}\right) +$

$\dfrac{1}{2}k2x\dfrac{dx}{dt} = 0$ and so canceling the dx/dt term $d^2x/dt^2 + kx/m = 0$.

12.63 $v = f\lambda$ multiply both sides by 2π to get $2\pi v = 2\pi f\lambda$, $(2\pi/\lambda)v = 2\pi f = \omega$.

12.71   Think of the wave as having been formed by raising a rectangular portion of water out, to form the trough, and up, to form the crest. That means an amount of work proportional to A, the height of the water was raised. To both deepen the trough and raise the crest by A requires an additional amount of work proportional to $3A$ giving a net proportional to $4A$ in order to create such a wave.

12.73

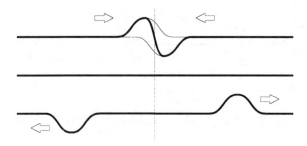

12.79 $\left(\dfrac{\partial x}{\partial t}\right)_{\varphi} = -\dfrac{\left(\dfrac{\partial \varphi}{\partial t}\right)_x}{\left(\dfrac{\partial \varphi}{\partial x}\right)_t} = \pm\dfrac{\omega}{k} = \pm\dfrac{2\pi f}{2\pi/\lambda} = \pm v_x$

12.83 $a_{max} = A(2\pi f)^2$; $v = \lambda f$; $f = v/\lambda = (2.0 \text{ m/s})/(1.0 \text{ m}) = 2.0$ Hz; hence $a_{max} =$
$(0.040 \text{ m})(2\pi\, 2.0\, s^{-1})^2 = 6.3 \text{ m/s}^2$.

12.87 $v = \sqrt{F_T/\mu} = \sqrt{(80 \text{ N})/(7.5 \times 10^{-3} \text{ kg/m})} = 1.0 \times 10^2 \text{ m/s}$.

12.93 $\Delta m = \Delta y(m/L)$; at height Δy, $F_T = g\Delta y(m/L)$ hence $v = \sqrt{F_T/(m/L)} = \sqrt{g\Delta y}$ the

speed increases as Δy increases. $V_{max} = \sqrt{gL}$. $F_T = g\Delta y(m/L) + Mg$ hence

$v = \sqrt{F_T/(m/L)} = \sqrt{g\Delta y + MgL/m}$.

Answers to Discussion Questions

13.1   The outgoing pulse will trigger a positive peak at the microphone 29 ms after launch, which will hit the open end 29 ms later, where it will be reflected with a 180° phase shift. Another 29 ms later the microphone will pick up a rarefaction and send out a negative peak. The scope will show a positive peak separated by 58 ms from a negative pulse.

13.3   The strings warm from the friction — they expand. There is a decrease in tension and a drop in speed and frequency. The wind instruments warm from the breath, increasing v and therefore f.

13.5   Explosions, firecrackers, gunshots, etc. all occur too rapidly for the acoustic reflex to be of much help. This is why everyone with any experience at an indoor pistol range will wear some sort of additional protection. The acoustic reflex works much better with loud sounds that build up gradually.

13.7   The triangular wave contains many strong overtones or Fourier components and so has a much greater high-frequency content, and sounds more "metallic." The sharper the bends in the string, the more high-frequency terms will be present. Plucking with the fingers will therefore generate fewer overtones and sound more mellow.

13.9   The answer lies in Young's Law of Strings (1800), which was discussed though not under its formal name. Remember that the point at which the string is struck must be an antinode because it is vibrating with maximum amplitude there. But the 7th harmonic has 7 antinodes and divides the string into 7 equal-length segments. Thus, 1/7th of the way down the string oscillating in the 7th harmonic, there must be a node. Hence, if we strike a string at that 1/7th distance, we preclude the presence of the unpleasant 7th harmonic.

13.13   The vibrating tuning fork transmits energy to the box, which is designed to have a standing wave pattern whose fundamental matches the frequency of the fork. The comparatively large box can impart energy to the surrounding air much more efficiently than could the fork whose vibrations have a tiny amplitude. Sound comes out especially effectively from the open end of the box. Evidently, the lower the frequency, the larger the sounding box.

13.17   Stroking the rod sets it into standing-wave vibration with a node at the midpoint support. The attached piston vibrates at that frequency and sets the air in the tube vibrating. Adjusting the plunger allows a standing wave to form in the doubly-closed tube and the powder is shaken from the air-displacement antinodes toward the nodes, where the air is still.

Answers to Multiple Choice Questions

1. a   3. d   5. e   7. d   9. d   11. b   13. c   15. a   17. b
19. a

Solutions to Problems

13.3 $v = (1500 \text{ Hz})\lambda$; $\lambda = (0.50 \text{ m/s})/(1500 \text{ s}^{-1}) = 3.3 \times 10^{-4}$ m.

13.7 $I = P/A = P/4\pi R^2 = (50 \text{ W})/4\pi(100 \text{ m}^2) = 4.0 \times 10^{-2}$ W/m². $E = IAt = (4.0 \times 10^{-2}$ W/m²$)(1.0 \times 10^{-4}$ m²$)(1.0 \text{ s}) = 4.0$ MJ.

13.17 $\Delta\beta = 10 \log_{10}1 = 0$ dB.

13.21   With the previous problem in mind $I = 10^{77/10}$ $(10^{-12}$ W/m²$) = 5.0 \times 10^{-5}$ W/m².

13.35   Since intensity varies as the square of the pressure $\beta = 10 \log_{10}(P/P_0)^2 = 20 \log_{10}(P/P_0)$.

13.47   The period of the beats is 0.99 s = $1/\Delta f$; $\Delta f = 1.0$ Hz.

13.51 $v = \sqrt{F_T/\mu} = 200$ m/s; $L = ½ \lambda$, $\lambda = 2.00$ m; $f_1 = 0.10$ kHz.

13.63   For $N = 1$, $y = 0$ (nodes) for all t when $x = 0$ and $x = L$. The antinode is at $x = L/2$. Remember that x is equal to or less than L. For $N = 2$, $y = 0$ when $x = 0$, $L/2$, and L, since for example, $\sin[2\pi(L/2)]/L = 0$. The antinodes are at $x = L/4$ and $3L/4$. In general $L = ½N\lambda$ and $N/L = 2/\lambda$; $v_y = -2\pi f_N A_N \sin kx \sin(2\pi f_N t)$

13.69   From Table 13.2, the ratio of speeds is (331.45 m/s)/(970 m/s) = (600 Hz)/f; $f = 1.76$ kHz.

13.83   From Eq. (13.17) $f_0 = (v + v_t)f_s/(v - v_t)$ hence the beat frequency is $f_0 - f_s = 2v_t f_s / (v - v_t)$.

Answers to Discussion Questions

14.1   Here are a few: (a) The bore hole in the glass tube must be uniform. (b) All of the mercury, including the stuff way up in the stem, must be at the same temperature. (c) Generally, to speed up its response the walls of the bulb are made thin and that thinness makes it vulnerable to pressure variations, which change its volume (via barometric changes or hydrostatic pressure, if it's immersed in a liquid). (d) There is a variation in pressure in the mercury due to the different heights of the column. (e) There is a difference in internal pressure if it's held vertically as opposed to horizontally. (f) There are errors associated with the softness of the glass. If the thermometer is raised to a high temperature and then cooled rapidly, it might take weeks for the glass to return to its original volume. Try measuring the freezing point of water before and immediately after reading its boiling point — the difference can be as great as 1°C. (g) The mere presence of the thermometer in a small system may change the temperature of the system. (h) When measuring a changing temperature at any moment, the thermometer will always read warmer if the bath temperature is falling and vice versa.

14.3   We know that the antimony expands on solidifying, as does water. Since that would be a very helpful trait for a casting material to have (it would fill all the fine details in the mold), it's reasonable to expect that's the answer to this question.

14.7   Being a poor conductor, the center of a boulder so treated would remain quite hot while the outside was cooled and contracted rapidly. Pressure would build up, there would be considerable internal stress, and the thing would rupture at any flaw or weak spot.

14.9   Bulbs are cheaper thinner and can tolerate the changes in temperature associated with ordinary operation, which are fairly gradual. They are made of inexpensive glass with a relatively large β, so a drop of water (or latex paint) can cause enough contraction and stress to shatter a hot light bulb. Clearly, outdoor lamps have to be protected from rain and snow. The heating in a flashbulb is so rapid even the thin walls tend to burst and they therefore are usually enclosed in a tough plastic film to keep them from shattering.

14.11   At a temperature of even -1°C it takes about 140 atm of pressure (≈ 2000 lb/in^2) to melt ice, so the problem was probably that the snow was just too cold.

14.13   The liquid will expand rapidly: its density decreasing as the density of the vapor increases. The surface meniscus will flatten out and then disappear altogether when the density of liquid and vapor are equal at a pressure of 7.38 MPa.

14.15   Yes. They were called permanent because they could not at first be liquefied, which suggests a weak intermolecular cohesive force and that, in turn, suggests ideal behavior.

14.17 $PV = nRT$, so both pressures must be equal since everything else is the same. The speeds of the hydrogen molecules must be greater than those of the nitrogen because the average

KE is the same from Eq. (14.14). The pressures can be equal because the lighter hydrogens hit the walls at greater speeds and they do it more frequently because they traverse the chamber more quickly. The pressure is proportional to the average KE via Eq.(14.13).

14.19   Remember that the pressure in the room is more or less constant. Increasing T increases the average KE, which increases the net KE and the P but leads to an over-pressure and an outward current of air. The room leaks warm air to the outside. The temperature goes up because it's dependent on the average KE of each molecule, which is higher. The pressure remains the same because it depends on both the number of molecules per unit volume and their average KE. Compare Eqs. (14.13) and (14.14).

Answers to Multiple Choice Questions

1. b	3. a	5. c	7. d	9. a	11. b	13. a	15. a	17. b
19. e	21. a							

Solutions to Problems

14.3 $0°F = -17.8°C$, $100°F = 37.8°C$ hence $\Delta T = 37.8 - (-17.8) = 55.6$ K.

14.11 $\Delta L = (25 \times 10^{-6}$ K$^{-1})(10$ m$)(20$ K$) = 5.0$ mm.

14.17 $PV = nRT$, $\beta = \dfrac{1}{V}\dfrac{dV}{dT} = \dfrac{1}{(nRT/P)}\dfrac{nR}{P} = \dfrac{1}{T}$

14.21 $\Delta V = \beta V_0 \Delta T = (182 \times 10^{-6}$ K$^{-1})(0.50 \times 10^{-6}$ m$^3)(100$ K$) = 0.009\ 1 \times 10^{-6}$ m^3. $V = 0.50$ cm$^3 + 0.009\ 1$ cm$^3 = 0.51$ cm^3.

14.31   Let τ be the period; $\tau = 2\pi\sqrt{L/g}$; $\tau = 2\pi\sqrt{(1.000\ \text{m})/g} = 2.006$ s. $\Delta L = \alpha L_0 \Delta T = (25 \times 10^{-6}$ K$^{-1})(1.000$ m$)(-20$ K$) = -0.50$ mm. Finally $L = 0.999\ 5$ m, and $\tau = 2.005\ 9$ s. The clock ran fast.

14.39 $P_i V_i / T_i = P_f V_f / T_f$; (99 kPa)(1200 cm^3)/(288 K) = (101.3 kPa). $V_f /(273$ K$)$; $V_f = 1.1 \times 10^3$ cm^3.

14.47   Each oxygen molecule has a mass of 32 u hence 16 g is ½ mol and 16.0 kg is 500 mol hence $(6.022 \times 10^{23}$ molecules/mol$) \times (500$ mol$) = 3.01 \times 10^{26}$ molecules.

14.55 $v_{rms} = \sqrt{3k_B T/m} = [3(1.380\,662 \times 10^{-23}$ J/K$)(293.15$ K$)/(5.313\,6 \times 10^{-26}$ kg$)]^{1/2} = 478.03$ m/s.

14.61   (a) The downward force is mg, the upward force is PA, and so $P = mg/A$.

(b) Displacing the piston by an amount Δy will change the volume and the pressure and therefore the force on the piston. Thus we need to look at dP and dV: $PV = C$; hence $P\,dV + V\,dP = 0$ and $dP = -P\,dV/V$ for a small displacement $\Delta P = -(mg/A)\Delta V/V = -(mg/A)A\Delta y/V = -mg\Delta y/V$; $A\Delta P = \Delta F = -mgA\Delta y/V$.

14.73   Begin with Eq. (14.12), in the form $P = Nm(V^2)_{av}/3V$; Nm is the total mass so $Nm/V = \rho$ hence $(3P/\rho)^{1/2} = \sqrt{(v^2)_{av}} = v_{rms}$.

14.79   The nitrogen expands from 1.0 liter to 4.0 liters and the pressure drops to $\frac{1}{4}$ 2.0 atm = 0.50 atm. The oxygen expands from 3.0 liters to 4.0 liters and its pressure ($P_iV_i = P_fV_f$) drops to $(3/4)5.0$ atm = 3.75 atm hence the total pressure is 0.5 atm + 3.75 atm = 4.25 atm = 4.3 atm.

Answers to Discussion Questions

15.3   Bicycling consumes the same amount of energy per hour as shivering but converts 80% of it into thermal energy. Shivering, which doesn't perform any work to speak of, converts chemical energy almost completely into thermal energy. We shiver when the body loses too much heat and needs a quick infusion of thermal energy.

15.5   Copper is a much better conductor than iron or steel, but iron is cheaper and chemically a better material to cook food on. The copper layer is there to rapidly and uniformly distribute the heat by conduction. To the same end, good pots are generally made thick-bottomed, but that's not necessary if one just wants to boil water.

15.7 Q/t should be proportional to the exposed area A, the temperature difference between the body and the fluid ΔT, and the physical parameters that describe the particular situation, such as geometry, orientation, wind speed, etc. All of the latter are combined into a constant of proportionality, K_c, called the convection coefficient, which is determined experimentally. Hence, $Q/t = K_c A \Delta T$. For a human on a windless day, $K_c \approx 5$ kcal/m^2·h·C°, and A is the exposed area of the skin.

15.9   We tend to be more comfortable in a dry-air environment when it's hot because the body can perspire and control its temperature more effectively when evaporation is rapid. The rate of evaporation decreases as the amount of water vapor in the air increases. When the body gets too warm, the blood vessels at the skin dilate so as to bring more blood to the surface which makes the skin red.

15.11   The ice forms first where it is in contact with the tray (which is a better conductor than the air) and then across the top, thereby encapsulating some water. Since ice is not a very good conductor, the remaining water takes a while to freeze. The warm water will be cooled more effectively, especially by evaporation, and enough of it may evaporate so that it will win the race. Even so, there are a lot of variables and the cool water often freezes first.

15.13   The red arrows indicate the rate of flow of heat that decreases in the exposed rod where there are losses (mostly via convection), and remains constant in the insulated rod where there are not. In the exposed rod, there is less and less thermal energy available and the flow diminishes. Correspondingly, the slope of the T-d curve, which is the *temperature gradient*, also diminishes. In the insulated rod, the heat flow is uniform and the temperature gradient constant; that is, the slope is constant. Heat is transported by the temperature gradient much as a liquid is propelled through a pipe by a potential energy gradient.

15.17   Wearing the hair inside will trap air that will not be disturbed by external winds. The low thermal conductivity of the air layer is what is important and that is better achieved with the fur inside. To stay cool, wear light-colored, porous, light-weight clothes. These will reflect radiant energy and facilitate evaporation of sweat. If there isn't much water to be had, evaporation of perspiration should be restrained as much as possible.

15.19   Thermal radiation from the Sun will strike a portion of your suit and it will absorb energy at a rate dependent upon its surrounding face characteristics. The intensity of that incident radiation is determined in part by the distance to the source — the farther away, the less power per unit area. Moreover, the suit will radiate over its entire surface at a rate dependent on its temperature (which, in turn, is partly determined by the body's output). The same sort of thing will happen to the thermometer. It will read an equilibrium temperature that is determined, among other things, by the fraction of its area illuminated. At the distance of the Earth from the Sun, a sphere in thermal equilibrium will stabilize at about 290 K, not far from room temperature. Thus, the thermometer reads its own temperature and not that of space.

Answers to Multiple Choice Questions

1. e   3. c   5. c   7. b   9. a   11. c   13. a   15. b   17. b
19. c

Solutions to Problems

15.7 $Q = cm\Delta T = (1.00 \text{ kcal/kg·K})(30 \text{ kg})\Delta T = 500 \text{ kcal}; \Delta T = 17 \text{ K}.$

15.15 $(100 \text{ g})(0.113)(1.00 \text{ cal/g·K})(50 \text{ K}) = m(1.00 \text{ cal/g·K})(5 \text{ K}); m = 113 \text{ g}.$

15.23 $540 \text{ ft·lb} = 732 \text{ J}; Q = 732 \text{ J} = c_L m_L \Delta T + c_w m_w \Delta T = \Delta T[(130 \text{ J/kg·K})(10.24 \times 10^{-3} \text{ kg}) + (1700 \text{ J/kg·K})(1.0 \text{ kg})] = \Delta T(1701 \text{ J/K}); \Delta T = +0.43 \text{ K}.$

15.27 $Q = (900 \text{ J/kg·K})(0.50 \text{ kg} + 0.60 \text{ kg})(12 \text{ K}) + (4186 \text{ J/kg·k})(2.00 \text{ kg})(12 \text{ K}) = 112\ 344 \text{ J} = 1.1 \times 10^5 \text{ J} = 27 \text{ kcal}.$

15.35   This requires an input of $Q = cm\Delta T = (4186 \text{ J/kg·K}) \times (1.00 \text{ kg})(100 \text{ K}) = 418\ 600 \text{ J}.$ The heat of combustion of hard coal is 33 MJ/kg hence we need $(0.418\ 6 \text{ MJ})/(33 \text{ MJ/kg}) = 0.012\ 7 \text{ kg} = 12.7 \text{ g}.$

15.75   The amount of heat needed to raise the water from 0°C to 80°C is the same as the amount needed to melt the ice at 0°C regardless of how much water and ice there is, provided they're equal. Hence the ice melts and we end up with twice as much water at 0°C.

15.81 $Q = c_w m_w \Delta T_w = (4186 \text{ J/kg·K})(0.900 \text{ kg})(95 \text{ K}) = 0.357\ 90 \text{ MJ} = 0.36 \text{ MJ} = c_s m_s \Delta T_s = (230 \text{ J/kg·K})(1.20 \text{ kg})(\Delta T_s); \Delta T_s = 1296.7 \text{ K}; T_f = 1575 \text{ K}.$ The melting point of silver is only 1234 K so it cannot get up to 1575 K without liquefying. Hence $Q = (230 \text{ J/kg·K})(1.20 \text{ kg})(960.8°C - 5.0°C) = 0.268\ 30 \text{ MJ}$ leaving 0.094 099 MJ. Melting 1 kg of silver requires $Q = mL_f = (1.00 \text{ kg})(109 \times 10^3 \text{ J/kg}) = 109 \text{ kJ}$ so it all doesn't melt and $T_f = 1233.95 \text{ K}.$

Answers to Discussion Questions

16.1   Air is drawn from $A-B$ as the piston moves out. From $B-C$, it is compressed adiabatically and rises in temperature. Fuel is sprayed in at C and immediately explodes because of the high temperature. $C-D$ is the isobaric combustion leg when heat enters. At D, the fuel is burnt and the piston continues to move outward. $D-E$ is the rapid and therefore adiabatic expansion where the hot gas does work. E is the end of the cycle and it's there that the exhaust valve opens. $E-B$ is an isovolumic drop to the outside pressure, accompanied by the loss of heat via the exhausting of hot gas. With the exhaust valve still open, the remaining gas is ejected as the piston moves inward from B to A.

16.3   Steam (blue arrows) enters behind the piston and drives it to the left as the slide valve moves to the right. Steam in the left half of the cylinder (red arrows) escapes through the exhaust port to the low-pressure condenser. The valve on the right then closes, the steam expands, and the piston moves left. Now the valve on the right opens to the exhaust, steam enters via the valve on the left and the piston moves right.

16.5   Warm, moist low-density air rises (via thermals) doing work on the atmosphere as it expands. Accordingly, since the process occurs quickly, and gases are very poor conductors, it will be essentially adiabatic. The internal energy of the uprush of air will decrease and its temperature will drop. Water will then condense out, forming a cloud.

16.7   The Sun provides the energy stored chemically in wood, coal, and oil. Order increases locally as the plants and animals storing solar energy grow, ultimately to form fossil fuels, but increase in the disorder of the Sun overbalances that.

16.9   The Sun evaporates water that comes down to the rivers and reservoirs as rain, thereafter to pour down a waterfall and drive a turbine. In effect, the Sun does work against gravity.

16.11   The process is adiabatic and so isentrophic. Work is done by the gas on the piston. Therefore, the internal energy decreases, and the temperature (T) drops. The volume obviously doubles and the pressure decreases. The entropy must stay constant since $Q = 0$.

16.13   The engine actually exhausts gas at a much higher temperature than 300 K. Fuel is incompletely burned, heat is conducted and radiated from the engine, and friction is present as well.

16.15   Yes it is possible, but with so many molecules the likelihood of all of them being in one corner is minute. Of course, if there were only 1 molecule flying around, the chance of "all the molecules" being in one corner would be fairly high. It's less likely with 10 molecules and much less still with 10^{28}.

Answers to Multiple Choice Questions

1. b   3. d   5. d   7. b   9. e   11. b   13. c   15. a   17. d
19. b

Solutions to Problems

16.7   From Eq. (16.2), $\Delta U = Q - W = 418.6\ J - (-100.4\ J) = 519\ J$.

16.15 $e_c = 1 - (T_L/T_H) = 1 - (293\ K)/(310\ K) = 5.5\%$.

16.23

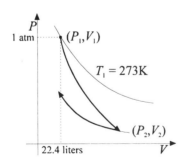

16.35   From Eq.(16.2), $\Delta U = Q - W$; $Q = 0$; $-\Delta U = W = P\Delta V = (99\ kPa)\cdot(4950 \times 10^{-6}\ m^3) =$
490 J. $Q = mL_v = 490\ J$; $m = (490\ J)/(2259\ kJ/kg) = 0.22\ g$.

16.39 $T_i = 273\ K$. From the diagram the lowest pressure is P_2; $P_f = P_i(V_i/V_f)^\gamma =$
$(1\ atm)(\frac{1}{2})^{1.4} = 0.379\ atm = 0.038\ 4\ MPa$. (b) $T_2 = P_2V_2T_i/P_iV_i = 206.9\ K = 207\ K$.

16.47   2000 lb → 907.2 kg; $Q_L = mL_f = (907.2\ kg)(333.7\ kJ/kg) = 302.7\ MJ$. $\eta =$
Q_L/W_i; $W_i = Q_L/\eta_c = (302.7\ MJ)/8.78 = 34.5\ MJ$ (per day), or 399 W.

16.51   From Eq.(16.16), $Q_L/W_i = \eta = T_L/(T_H - T_L)$; W_i(per minute) = (360 J/min) ×
(31/269) = 41.49 J/min = 0.69 W.

16.55   From Eq. (16.11), $e = 1 - 450/1200 = 62.5\%$; as an engine it had $Q_H = 1200$ J, $Q_L = 450$ J, and $W = 750$ J and those numbers must be the same if the engine is reversible even if now heat-out = 1200 J rather than heat-in = 1200 J, etc. But here work-in is and not 750 J so it's not reversible.

16.59 $Q_H = ST_H = (30$ kJ/K$)(300$ K$) = 9.0$ MJ; $Q_L = -(30$ kJ/K$) \times (100$ K$) = -3.0$ MJ. $W = 9.0$ MJ $- 3.0$ MJ $= 6.0$ MJ; $e = (6.0$ MJ$)/(9.0$ MJ$) = 66.7\% = 67\%$.

16.63   (a) The internal energy of the gas of N molecules is all KE; $U = 3Nk_BT/2$ and so a change in temperature produces a change in internal energy $dU = 3Nk_B\,dT/2$; (b) $PV = Nk_BT$ and $dW = P\,dV = [(Nk_BT)/V]\,dV$; $dU + dW = 0$ and so $(dT/T) + (2/3)(dV/V) = 0$; $\ln T + (2/3)\ln V = $ constant; $TV^{2/3} = $ constant; $PV = $ constant $\times T$; $PV^{5/3} = $ constant.

16.67 $dW = 0$; $dQ = 0$; $T_f = T_i$; to determine the entropy change we need to use a reversible process between state-i and state-f at a fixed temperature; since it's isothermal, $dU = 0$; $dQ = dW = $

$$P\,dV \text{ and } S = \int \frac{dQ}{T} = \frac{1}{T}\int_{V_i}^{V_f} P\,dV = \frac{nRT}{T}\int_{V_i}^{V_f}\frac{dV}{V} = nR\ln\frac{V_f}{V_i};$$ since $V_f > V_i$, the change in entropy is

positive.

Answers to Discussion Questions

17.1   The paper is electrically polarized by the field, and is therefore attracted to the charged conductor (or sheet of glass). It is only after a little while that charge is transferred to it from the highly charged object. At that point, the paper has a net charge with the same polarity as the conductor (or glass) and is repelled.

17.3   The field is strongest in the region between the charges. The field lines are perpendicular to the walls and because the chamber is grounded there is no outer-surface charge and no field outside.

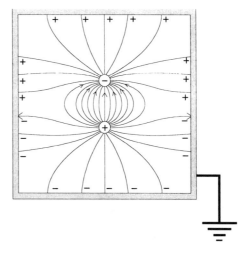

17.7   Rubbing the tube electrified it and the free charges flowed along the moistened string, which was a fairly good conductor. The lead ball became charged as if it had been touched directly by the glass rod. It then polarized the leaf, which was attracted upward to it.

17.9   As we will discuss later, the typewriter radiates an electromagnetic wave that the antenna wire picks up and superimposes on the signal coming down from the roof. One thing to do is to wrap the antenna wire, in the vicinity of the typewriter, with aluminum foil and ground it, thus shielding it. Another (though we will not discuss it until later), is to reorient the portion of the wire near the typewriter so it is not along the radiated E-field.

17.15   *The case of attraction is always ambiguous* — the elephant might be either negative or neutral, whereupon the positive ball would induce a negative charge and be attracted. The repulsion unambiguously means that the elephant was negative.

Answers to Multiple Choice Questions

1. c   3. b   5. c   7. a   9. a   11. d   13. b   15. c   17. d
19. e

Solutions to Problems

17.11   The net force is $2(F_{31} \cos \theta)$ since the horizontal components cancel; hence $F = 2(kqq/r^2) \cos \theta$; $\theta = 45°$; $F = 2 \cos \theta \times (9.0 \times 10^9 \text{ N·m}^2/\text{C}^2)(+25 \times 10^{-9} \text{ C})^2/(1.0 \text{ m})^2 = 8.0 \ \mu\text{N}$, upward, along the perpendicular bisector of the hypotenuse.

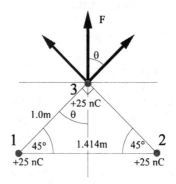

17.21 $F_{12} = kq_1q_2/r_{12}^2 = 3.6$ N;

$F_{13} = kq_1q_3/r_{13}^2 = 4.5$ N;

$F_{14} = kq_1q_4/r_{14}^2 = -1.8$ N;
$F_{x1} = (-1.8 \text{ N}) + (4.5 \text{ N})4/5 = 1.8$ N; $F_{y1} = -(3.6 \text{ N}) - (4.5 \text{ N})3/5 = -6.3$ N; $F_1 = \sqrt{1.8^2 + 6.3^2}$ N $= 6.6$ N; $\theta = \tan^{-1}(6.3/1.8) = 74°$.

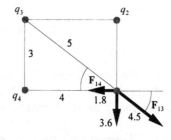

17.39   E at each vertex has a net component only along a line bisecting that angle (and pointing away from the triangle). $E = 2(kq/r^2) \cos 30° = 2(9.0 \times 10^9 \text{ N·m}^2/\text{C}^2)(20 \times 10^{-6} \text{ C})(0.866)/(2.0 \text{ m})^2$ and $E = 7.8 \times 10^4$ N/C.

17.45   They should be charged with $+q$ on the bottom plate and $-q$ on the top where $E = \sigma/\epsilon_0 = q/A\epsilon_0$; $q = (1000 \text{ N/C})(1.00 \text{ m} \times 0.50 \text{ m})(8.85 \times 10^{-12} \text{ C}^2/\text{N·m}^2) = 4.4$ nC.

17.55 $a = q_eE/m_e = (1.60 \times 10^{-19} \text{ C})(1.5 \times 10^4 \text{ N/C})/(9.11 \times 10^{-31} \text{ kg}) = 2.6 \times 10^{15}$ m/s^2.

17.61 $dq = \sigma dA = \sigma 2\pi r \, dr$; from Example 17.7 we know the field of a differential ring of radius r is given by $dE_x = \dfrac{kx \, dq}{(x^2 + R^2)^{3/2}} = \dfrac{kx(\sigma 2\pi r \, dr)}{(x^2 + R^2)^{3/2}}$; Using the fact that $2r \, dr = d(r^2)$ we make

r^2 the variable and get $E_x = \pi k\sigma x \int_0^R \frac{d(r^2)}{(x^2 + r^2)^{3/2}} = \pi k\sigma x \left[-\frac{2}{(x^2 + r^2)^{1/2}} \right]_0^R = 2\pi k\sigma \left[1 - \frac{x}{(x^2 + R^2)^{1/2}} \right]$

17.67   Draw a spherical Gaussian surface inside the ball at a radius r. By symmetry the E-field must be radial. $\rho = Q/V = Q/(4\pi R^3/3)$; the charge inside the surface is $\rho(4\pi r^3/3)$ hence $E(4\pi r^2) = \rho(4\pi r^3/3)/\epsilon_0$; and $E = r\rho/3\epsilon_0 = rQ/4\pi\epsilon_0 R^3$; the field increases linearly with distance inside the charge distribution.

Answers to Discussion Questions

18.3   The neutral conductor becomes polarized with a negative induced charge near the original sphere and a positive charge on the far side. These charges, in turn, contribute to the potential, with the negative charges dominating and reducing the potential at the original sphere. Accordingly, the potential of the original sphere (that is, inside) is lower. Outside, it drops rapidly to a finite value that is constant across the neutral conductor.

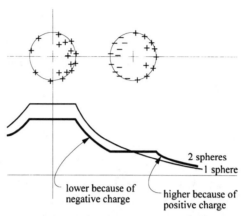

18.5   Since no charge flows from or to the neutral conductor while touching the inside wall, the two must be at the same potential, which is uniform in the cavity, implying that the conductor assumes the potential of the field-free region. In a region where there is a field, the conductor will distort the field pattern and change the potential. If more charge is added to the outer conductor, the potential inside increases and the potential of the neutral body increases.

18.9   The field lines go from the positive charges to the negative ones. There is no field at the very center. The equipotentials surround each of the charges, being negative in the vicinity of the negative charge. There are two zero equipotential planes that pass through the center point.

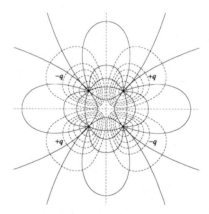

18.11   Bringing a positive charge to a point in space introduces an outwardly directed field and a positive potential everywhere in the region; a negative charge does just the opposite, contributing a negative potential and so lowering the net potential everywhere. Imagine a gold-leaf electroscope attached to a positively charged plate. The electroscope reads the potential of the plate. Now bring a positive charge near the plate; positive charge from the plate will be repelled back to the scope, the leaves will part even more, and it will thereby indicate an increase in potential.

18.15   There cannot be a charge since that would produce a field and therefore a potential gradient, but the potential is constant.

18.19   The electric field lines are perpendicular to the grid of curved equipotentials. The field lines, which become fairly straight, converge to point P (from the anode on the right) and then diverge away from P (on the left). Electrons in the beam approaching from the left are consequently accelerated toward P.

Answers to Multiple Choice Questions

    1. b   3. c   5. b   7. d   9. d   11. d   13. c   15. d   17. b
19. b

Solutions to Problems

18.3 $\Delta V = \pm Ed = \pm(1.00 \text{ V/m})(0.10 \text{ m}) = \pm 0.10 \text{ V}$.

18.9 $|\Delta PE_E| = |q_e\Delta V| = (1 \text{ electron})(500 \text{ V}) = \Delta KE = 500 \text{ eV}$. It loses 500 eV of PE_E.

18.29 $Ed = \Delta V, E = \Delta V/d = (85 \text{ mV})/(8.0 \text{ nm}) = 11 \text{ MV/m}$.

18.37 $E_x = -\dfrac{\partial V}{\partial x} = -kp\dfrac{\partial}{\partial x}\left[\dfrac{x}{(x^2 + y^2)^{3/2}}\right] = -kp\left[\dfrac{1}{(x^2 + y^2)^{3/2}} - \dfrac{3x^2}{(x^2 + y^2)^{5/2}}\right] =$

$-kp\dfrac{(y^2 - 2x^2)}{(x^2 + y^2)^{5/2}}$; similarly $E_y = -\dfrac{\partial V}{\partial y} = kp\dfrac{3xy}{(x^2 + y^2)^{5/2}}$.

18.43   The voltage across the capacitor is so high that it can be quite dangerous. $Q = CV = (500 \text{ pF})(20 \text{ kV}) = 1 \times 10^{-5} \text{ C}$, which is a lot of charge.

18.49 $C = \epsilon A/d = 10(8.85 \times 10^{-12} \text{ C}^2/\text{N·m}^2)(100 \times 10^{-4} \text{ m}^2)/(1.0 \times 10^{-3} \text{ m}) = 8.9 \times 10^{-10} \text{ F}$.

18.55   They are all in parallel; $C = 100(4.1\epsilon_0)A/d = 0.99 \text{ μF}$.

18.67   They are all in parallel, hence $C = 9 \text{ pF}$.

18.75   From Problem 18.71 we have that $C = 30 \text{ pF}$, hence $PE_E = \frac{1}{2}CV^2 = \frac{1}{2}(30 \text{ pF})(12 \text{ V})^2 = 2.2 \text{ nJ}$.

18.79   The system is actually two capacitors in parallel with the center sheet serving as a plate for both. For each capacitor $C = \epsilon A/d = (4.5)(8.85 \times 10^{-12} \text{ C}^2/\text{N·m}^2)(0.30 \text{ m}^2)/(1.0 \times 10^{-3} \text{ m}) = 12 \text{ nF}$, and so the net capacitance is 24 nF.

Answers to Discussion Questions

19.1   The flow of heat along a metal rod is proportional to the temperature difference across its ends, just as the flow of charge is proportional to the voltage difference. In metals, free electrons serve to transport both electrical and thermal currents.

19.5   This arrangement is for controlling a light bulb from two locations. Either switch will turn the bulb on or off independent of the other. As shown in Fig. Q5, the lamp is on. Throwing either switch open-circuits the bulb, shutting it off (as indicated here).

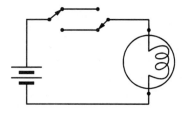

19.7   The gasoline engine powers a separate electrical generator that operates all systems and sends its excess current to the battery to recharge it. Once the engine is running, the battery can be completely removed from the circuit. Leaving off all unnecessary electrical devices will allow more current to be provided to recharge the battery.

19.11   The ammonium chloride dissociates into NH^+ (which, while current is circulating, migrates to the carbon rod) and Cl^-. Chlorine ions combine with the zinc electrode to make zinc chloride. If it were really dry, there would be no migration of the ions and it wouldn't work. Until only a few decades ago, the cell's outer casing was the thin zinc can, simply wrapped in a cardboard insulating sleeve. More often than not, the zinc would be eaten away and the ammonium chloride paste would leak out and corrode your flashlight. Even with today's steel casings, it's still not wise to store electronic equipment with batteries inside them for long periods.

19.15   With current fanning out radially from the point of impact, there will be a voltage drop across the animal, and if its resistance is not very much greater than the ground's, a sizeable current will pass through it. Squat.

Answers to Multiple Choice Questions

1. c   3. d   5. c   7. e   9. b   11. b   13. a   15. c   17. d
19. c   21. a

Solutions to Problems

19.5 $I = \Delta Q/\Delta t$; $\Delta Q = (180 \text{ A})(2.0 \text{ s}) = 3.6 \times 10^2$ C.

19.19 $\quad I = \dfrac{dq}{dt} = 3(4.00 \text{ C/s}^3)t^2 - (4.00 \text{ C/s})$ and therefore

$J = \dfrac{(12.0 \text{ C/s}^3)t^2 - (4.00 \text{ C/s})}{2.00 \times 10^{-6} \text{ m}^2} = (6.00 \times 10^6 \text{ C/s}^3\cdot\text{m}^2)t^2 - (2.00 \times 10^6 \text{ C/s}\cdot\text{m}^2)$. At $t = 1.00$ s,
$J = 4.00 \times 10^6 \text{ A/m}^2$.

19.23 $\quad$ (a) A and B, (b) A and E, (c) A, D and C, (d) A and D, (e) A and F. When the headlights are on.

19.31 $\quad \Delta Q$ is the area under the I versus t curve which is $\frac{1}{2}(6.0 \text{ h} \times 60 \text{ min/h} \times 60 \text{ s/min}) \times (7.0 \text{ A} - 3.0 \text{ A}) + (3.0 \text{ A})(21\,600 \text{ s}) = 43\,200 \text{ C} + 64\,800 \text{ C} = 0.11 \text{ MC}$.

19.35 $\quad V = IR$; $R = V/I = (100 \text{ kV})/(2.0 \text{ }\mu\text{A}) = 50 \times 10^9 \text{ }\Omega$.

19.43 $\quad A = \rho L/R = (1.7 \times 10^{-8} \text{ }\Omega\cdot\text{m})(1609.3 \text{ m})/(10 \text{ }\Omega) = 2.736 \times 10^{-6} \text{ m}^2$; $r^2 = 8.71 \times 10^{-7} \text{ m}^2$; $d = 1.9 \text{ mm}$.

19.53 $\quad$ P $= IV = (180 \text{ A})(12 \text{ V}) = 2.2 \text{ kW}$.

19.65 $\quad$ Current per strand $I_s = 0.012 \text{ A}$; $V_s = I_s R = 2.4 \times 10^{-5} \text{ V}$; (b) across the wire $V = 2.4 \times 10^{-5} \text{ V}$; (a) $R = V/I = (2.4 \times 10^{-5} \text{ V})/(0.12 \text{ A}) = 2.0 \times 10^{-4} \text{ }\Omega$; (c) $I_s^2 R_s = 2.9 \times 10^{-7} \text{ W}$.

19.69 $\quad$ P $= IV = [(10 \text{ }\mu\text{C})/(1 \text{ s})](3 \text{ MV}) = 30 \text{ W} = 0.03 \text{ kW}$.

19.77 $\quad \Delta R/R_0 = \alpha_0 \Delta T = (0.004\,5 \text{ K}^{-1})(600 \text{ K}) = 2.7$; clearly α_0 is not constant and the calculation is quite poor. At 2500° C α_0 can be crudely approximated graphically to be about 0.02 K^{-1}; $\Delta R/R_0 = \alpha_0 \Delta T \approx (0.02 \text{ K}^{-1})(600 \text{ K}) \approx 12$.

Answers to Discussion Questions

20.3   This circuit was supposed to simply illustrate the Node Rule at G; 12 amps flow in, 12 amps flow out. But they overlooked something. A-H-G and B-H-G are short circuits; A and G, and B and G are all at the same potential. That means that no current flows through the resistors in branches A-G and B-G, and the diagram is incorrect as labeled.

20.5   The switch is closed and the capacitor charges up slowly (depending on RC). As it does, the voltage across it increases until the breakdown voltage of the lamp is reached, at which point the capacitor rapidly discharges through the lamp, which flashes. The voltage across the capacitor, and therefore the lamp as well, drops rapidly only to slowly build up again for a repeat performance.

20.7   The circuit first sees the edge of the pulse and a voltage of V. Charge and voltage both gradually build on C until the signal drops to 0 at which point they both decay. If the time constant is short enough, V_C will reach nearly 0 before the next pulse arrives. Since the drop across the capacitor is initially zero, V_R starts at a maximum and decays exponentially to 0.

When the input signal drops to 0, it's as if the input terminals were shorted. The voltage in the circuit comes from the charged capacitor, and the resistor is across its terminals. When the upper input circuit terminal was positive, the current circulated clockwise. When the input is shorted, current (discharged from the capacitor) circulates counterclockwise. Hence, the voltage polarity on the resistor is reversed and decays from there to zero as the capacitor discharges. In general, $V_R = V - V_C$.

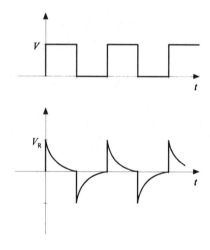

20.9   The benchmark of *direct current* is that it does not change direction, which means that a dc voltage must not change polarity or sign. It certainly can vary — it need not be constant — but it cannot reverse itself. a, c, d, e, g, and h are all dc. b and f are known as alternating current, ac.

20.11   1 is brightest; 2 and 3 are equally bright, though less so than 1; no current passes through 4, 5, or 6, and so they are off altogether.

20.13   The resistance of the middle branch is half that of the upper branch; it draws twice the current and so 3 is brighter than 1 or 2, which have the same current and are equally dim.

20.15 R_1 and R_5 have the most current; R_2, R_3 and R_4 have less; R_6 and R_7 have none. (R_7 has no voltage across it and there is no closed circuit through R_6).

Answers to Multiple Choice Questions

1. d	3. d	5. c	7. c	9. a	11. b	13. d	15. c	17. a
19. b	21. b							

Solutions to Problems

20.1 $R_e = 20\ \Omega$, $I = V/R_e = 0.60$ A before; after $I = V/R_e = (12\text{ V})/(8.0\ \Omega) = 1.5$ A. It's short-circuited and goes out.

20.13 $2\ \Omega$, $3\ \Omega$, and $6\ \Omega$ in parallel equals $1\ \Omega$ in parallel with $1\ \Omega$ equals $\frac{1}{2}\ \Omega$ in series with $5\ \Omega$ is $5.5\ \Omega$.

20.23 $P = IV$, 80 W $= I(20$ V$)$, $I = 4.0$ A; $V = IR$, $R = (60$ V $- 20$ V$)/(4.0$ A$) = 10\ \Omega$.

20.27 $RC = (5.0\text{ k}\Omega)(800\ \mu\text{F}) = 4.0$ s.

20.33   9 A through 17 Ω; 9 A splits, 1 part out of 10 or 0.1(9 A) = 0.9 A through 9 Ω, and (9/10)(9 A) = 8.1 A through the 1 Ω; 9 A enters and leaves the next node; 3 Ω + 1 Ω = 4 Ω, 4 Ω in parallel with 2 Ω in parallel with 5 Ω yields Re = 1.05 Ω: $V = (9$ A$)(1.05\ \Omega) =$ 9.47 V hence current through the 3 Ω and 1 Ω resistors is 2.4 A or 2 A. Through the 2 Ω passes 4.7 A or 5 A. Through the 4 Ω and 1 Ω resistors pass 1.9 A or 2 A.

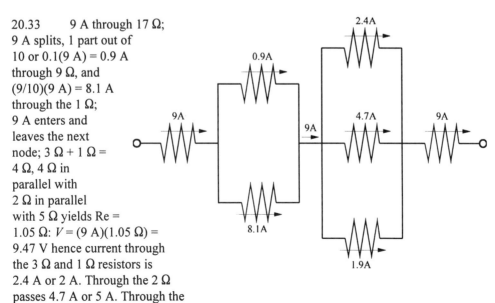

20.53   The voltage across A-D-C is 12 V hence the current is (12 V)/(4 Ω) = 3 A. The voltage across the 2-Ω resistor in loop A-B-C-A is 12 V - 6 V hence the current through it is 3 A and from the Node Rule the current through the 12-V battery is 6 A.

20.65   The voltage between the top and bottom nodes is 48 V - (3 A)(4 Ω) = +36 V. Hence, R_1 = 46 Ω and $\mathcal{E}_1$ = 14 V.

20.71 $I_2 = I_6 = 2.0$ A, $I_4 = I_5 = 5.0$ A, $V = (10\ \Omega + 5.0\ \Omega)(2$ A$) = 30$ V $= (2.0\ \Omega + R)(5.0$ A$)$; $R = 4.0\ \Omega$. $I_3 = 1.0$ A, $I_1 = 8.0$ A.

Answers to Discussion Questions

21.1   The electrostatic E-field of a point charge should have spherical symmetry because the charge presumably has spherical symmetry and space is isotropic. Once the charge is set moving, the axis of that motion represents a distinguishable direction and both the E and B fields will now only be axially symmetric (one circular, one radial).

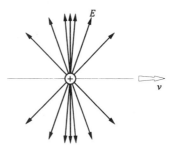

21.3   Each north pole would experience a torque of $q_m B_N R_N$ and each south pole a torque of $q_m B_S R_S$ in the opposite direction. These must cancel since there's no rotation, hence they are equal and $B_S/B_N = R_N/R_S$; the field varies inversely with the radial distance.

21.7   The field of the coil rapidly changes magnitude and direction and that tends to mix up the domain structure. Since $B \to 0$ as the coil is removed, the domains cannot continue to realign themselves with the field of the coil.

21.11   Far from any long wiggly wire the field must resemble that of a straight wire. Certainly if the bends are perfectly regular (e.g., an equilateral sawtooth pattern or a sinusoid) we could expect the same field (some distance away) as from a straight current. Since field lines do not cross we can imagine sets of circular lines around each tiny segment of the bent wire. These can be thought of as flattening out into planes perpendicular to the axis of the wire at distances large compared to the wiggles — remember the E-field of two equal charges. Ampère's Law makes no distinction between wiggly currents and straight ones. Experimentally, it is found that the setup will be unaffected by the introduction of a B-field. A solenoid is a spiralling wire, and if it has a projected length (end-to-end), it will have the external field of a straight wire.

21.15   Looking down on the left side, there is a counter-clockwise B-field and the north pole rotates counterclockwise around with the field. On the right, there is a downward current in the B-field of the magnet, and therefore the hanging rod experiences a perpendicular force that causes it to rotate clockwise.

21.17   The magnet's field was expelled from the disk when the latter went superconducting. The result is very much as if the field lines were bent up and away by the presence of an identical magnet (with like poles) within the superconductor that repels the cube.

Answers to Multiple Choice Questions

1. d   3. c   5. c   7. b   9. b   11. a   13. a   15. b   17. c
19. c   21. a

Solutions to Problems

21.1   From Eq. (21.7) 1 T = 1 N/(C·m/s); 1 T·m/A = 1 (N·s/C·m)·m/A = 1 (N/A)/A = 1 N/A².

21.13 $B_z \approx \mu_0 nI$; $I = 0.20$ A.

$\mathbf{v} = d\mathbf{l}/dt$, $I\,d\mathbf{l} = (Q/dt)\mathbf{v}\,dt$; $dB \rightarrow B$ and so since $\hat{\mathbf{r}} = \mathbf{r}/r$; $\mathbf{B} = \dfrac{\mu_0}{4\pi}Q\dfrac{\mathbf{v} \times \mathbf{r}}{r^3}$.

21.19 $I = (6.0 \times 10^{12}\ \text{s}^{-1})(1.6 \times 10^{-19}\ \text{C}) = 9.61 \times 10^{-7}$ A; $B = \mu_0 I/2\pi r = 1.3 \times 10^{-11}$ T, clockwise looking toward the source.

21.27   At $z = 0$ it becomes Eq. (21.3). As z gets very large ($z \gg R$) R is negligible in the numerator and the equation becomes $B_z = \dfrac{\mu_0 IR^2}{2z^3}$

21.33   Using Ampère's Law for a circular path in the plane of the torus and lying within the central hole ($r < R_i$), the field must be circular if it exists at all. But no current is encompassed so $B\Sigma\Delta l = \mu\Sigma I = 0$ and $B = 0$. The same would be true outside the torus where $r > R_0$. There the same number of wires carry current into the plane of the Ampèrean path loop as out of that plane. Hence $\Sigma I = 0$ and $B = 0$.

21.43   Only two segments contribute to B, the straight wire of length L and the curved wire. Thus from the previous problem the straight piece contributes $B_s = \dfrac{\mu_0}{4\pi}\dfrac{IL}{h\sqrt{L^2 + h^2}}$; while the

curved piece (from Problem 41) contributes $dB_c = \dfrac{\mu_0}{4\pi}\dfrac{I\,dl\sin\theta}{R^2} = \dfrac{\mu_0}{4\pi}\dfrac{IR\,d\varphi}{R^2} = \dfrac{\mu_0}{4\pi}\dfrac{I\,d\varphi}{R}$;

and $B_c = \dfrac{\mu_0}{4\pi}\dfrac{I(\pi/6)}{R} = \dfrac{\mu_0}{24}\dfrac{I}{R}$; $B = B_s - B_c = \dfrac{\mu_0}{4\pi}\dfrac{IL}{h\sqrt{L^2 + h^2}} - \dfrac{\mu_0}{24}\dfrac{I}{R}$.

21.47   From Problem 21.35 overlapping the two fields from the sheets it's clear that $B = 0$ outside and $B = 2(\frac{1}{2}\mu_0 i)$ inside. Alternatively, using Ampère's Law with a rectangular loop of width l with one side between the sheets, one above and the other two perpendicular to the sheets: $\Sigma B_{\parallel}\Delta l = \mu_0 \Sigma I$; $Bl + 0 + 0 + 0 + 0 = \mu_0 il$; and $B = \mu_0 i$, here B is perpendicular to two sides and zero over the third.

21.59 $\mathbf{F} = q\mathbf{v} \times \mathbf{B}$ the force is in the $-y$ direction.

21.79 $F_M/l = \mu_0 I_1 I_2/2\pi d = 2.0 \times 10^{-5}$ N, repulsive.

21.81 The force is everywhere perpendicular to the table. $dF = I(dl)B\sin\theta$; let $dl = dx$;

$B = C/r^2 = C/(h^2+x^2)$ and $\sin\theta = h/r = h/\sqrt{h^2 + x^2}$; $F = \int I(dl)B\sin\theta = hIC \int\limits_{-L/2}^{+L/2} \dfrac{dx}{(h^2 + x^2)^{3/2}} =$

$hIC \left[\dfrac{x}{h^2\sqrt{x^2 + h^2}} \right]_{-L/2}^{L/2} = \dfrac{IC}{h} \left[\dfrac{L}{\sqrt{(L/2)^2 + h^2}} \right]$

Answers to Discussion Questions

22.3   The handle of the device slips into a coil hidden in the base of the holder. Another coil is inside the handle and the two are in close proximity. Connecting the coil in the well to AC generates a time-varying B-field that passes through the plastic skin of the handle and produces a rapidly varying (60 Hz) induced emf and an induced current. That current is converted to DC and used to charge a battery in the handle, which powers the motor in the device. No charge and so no "electricity" passes from the base to the handle, although electrical energy certainly is transferred. The same scheme will work across human skin to power a mechanical heart.

22.5   As the magnet approaches, its B-field attempts to penetrate the ring. A supercurrent is thereby induced opposing the buildup of flux and the motion of the magnet. If the ring is free to move, it will lift off its support and hover (with its induced north pole down) above the north pole of the magnet. Finite, persistent supercurrents *on the surface* of the superconductor circulate in such a way as to shield the interior from the field. No flux enters the body of the superconductor; there is no change in flux, no E-field induced, and no bulk current.

22.7   To crank the generator without a load and therefore without a current (in the steady state), one need only overcome friction. By contrast, lighting the 100-W bulb requires twice the power used in lighting the 50-W bulb and that power must be supplied by the person turning the crank. Most people will find it difficult to keep the 100-W bulb lit very long.

22.9   Current would be induced in the loop and power transferred from the line to it — the time-varying B-field would induce an E-field. If a wire loop was in that region, the E-field would do work on the free electrons, transferring energy to them. The power company could detect a loss in energy arriving at the end of the line; there would be a slight drop in delivered current since the voltage is fixed by the generator. If you separate the two leads from a telephone and lay the pick-up loop (attached to a good set of headphones) next to one of the wires, you should be able to listen in on the time-varying B-field of the conversation.

22.13   Yes, a coil in a motor turning through a magnetic field will experience an induced emf and an induced current that will send energy back to the source, which is driving the motor. With no load, the motor produces (and returns to the power company) almost as much current as it draws, and so costs very little to run. A free-turning motor must have its speed limited by the back-current it generates — with no losses (in the bearings, etc.) the motor will speed up until the back-current equals the driving current and it can no longer accelerate. With a load, the motor does work and draws energy in excess of what it returns via the back-current. In real life, a motor also produces a good deal of thermal energy via friction and if it is to operate continuously for any long period of time, it must be cooled (usually by forced air). A jammed motor will not turn and not generate a back-current. The driving current (which depends on the motor's resistance) now undiminished by any back-current is too great. The wiring will heat up and the insulation will begin to burn off. If the process is not stopped soon, the motor will be destroyed.

22.15   Recall that the B-field outside a very long tight solenoid approaches zero. Each meter reads the potential drop across the resistor adjacent to it, the resistor with which it forms a closed circuit excluding the solenoid. The induced emf equals the difference between the two meter readings, or alternatively, the sum of the potential differences across the resistors in the central circuit (1-9-10-2-7-8-1). The seemingly strange thing here is that the "voltage" measured between 1 and 2 actually depends on how you hook up the meter. The varying flux provides energy to the induced current (energy coming from the power source driving current through the solenoid). In part (b) both meters would read $I_i R_2$.

Answers to Multiple Choice Questions

| 1. e | 3. c | 5. a | 7. a | 9. c | 11. c | 13. b | 15. d | 17. d |
| 19. b | 21. b | | | | | | | |

Solutions to Problems

22.3 $\Phi_M = BA$; $B = (6.0 \text{ mWb})/(0.005\ 0 \text{ m}^2) = 1.2 \text{ Wb/m}^2$.

22.7   From Eq. (22.3), emf $= -(1)(0.25 \text{ m}^2)(-0.40 \text{ T})/(0.200 \text{ s}) = 0.50 \text{ V}$.

22.15   The bulb would not light just as a voltmeter across the tips would not indicate a voltage difference. The meter and its leads would have the same voltage induced across them and once attached to the wing no current would circulate. In other words, the flux through the closed loop of meter-leads-wings does not change if the field does not change. You could read a voltage difference if the meter stayed on the ground and still managed to remain connected to the plane.

22.19   At any instant $\Phi_M = B\pi R^2 \cos\theta = B\pi R^2 \cos\omega t$; $\dfrac{d\Phi_M}{dt} = -\pi R^2 \omega B \sin\omega t$; $\mathscr{E} = \pi R^2 \omega B \sin\omega t$.

22.27 $\mathscr{E} = 220(20 \text{ mT/s})\pi(0.10)^2 = 0.138 \text{ V}$; $CV = Q = 4.1\ \mu\text{C}$.

22.35 $\Delta\Phi_M = 2BA$; $B = QR/2NA = 0.83 \text{ mT}$.

22.41 $A = \dfrac{1}{2}L(L\theta) = \dfrac{1}{2}\omega t L^2$; $\mathscr{E} = -\dfrac{d\Phi_M}{dt} = -B\dfrac{d}{dt}\left(\dfrac{1}{2}\omega t L^2\right) = -\dfrac{B\omega}{2}L^2$. The minus sign just corresponds to Lenz's Law. The flux change equals the number of lines cut, so the analysis is correct from that perspective.

22.49   The maximum voltage is 100 V. $\omega = 2\pi f = 376.99$, $f = 60.000$ Hz.

22.67   Across R, $V = IR = (1.25 \text{ mA})(20\ \Omega) = 25 \text{ mV} = $ induced emf. $\mathscr{E} = \frac{1}{2}Br^2\omega = \frac{1}{2}Br^2 2\pi f$;

$B = (25 \text{ mV})/(0.25 \text{ m})^2 \pi(6 \text{ Hz}) = 21 \text{ mT}$.

22.71   We want $\mathscr{E} = NAB\omega \sin \omega t$; 6000 rpm = 100 rps, $\omega = 2\pi(100 \text{s}^{-1})$; $NAB\omega = (200) \times$

22.75 $L/l \approx \mu_0 N^2 A/l^2 = 6.32 \text{ mH/m}$; $PE_M = \frac{1}{2}LI^2$; energy per unit length is 79 mJ/m.

22.79   (a) $I_m = V/R = 2.0 \text{ A}$; (b) $L/R = 4.0 \text{ s}$; (c) $\approx 5L/R \approx 20 \text{ s}$.

Answers to Discussion Questions

23.3   Unlike ac, dc does not induce currents in nearby conductors, thereby reducing the transmitted power. dc can be transmitted at higher voltages since it does not peak. The same *rms* voltage for an ac signal will rise to a peak voltage (1.414 times higher) and be more troublesome as far as insulation and breakdown of air is concerned).

23.7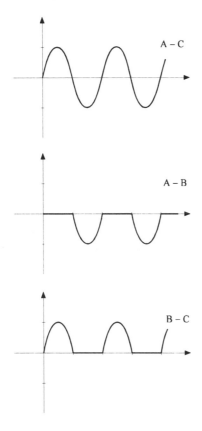

Across A-C, the scope reads the sinusoidal voltage induced on the output side of the transformer. Across A-B, the scope reads the voltage across the diode, which is zero when current is flowing through (when the diode acts like a closed switch) and equals the transformer output voltage when no current is flowing (when the diode acts like an open switch). Across B-C, the meter reads the voltage across the resistor, which equals the transformer output voltage when current is flowing and equals zero when no current is flowing.

23.9   Put the tester across the terminals of the fuse holder without the fuse in place — if all is well, the side coming into the house will be hot, the other will be floating, there will be a sizeable potential difference and the neon will light. Now screw in the fuse and repeat the last test. If the fuse is good and if it is properly seated in the holder, there will be no appreciable voltage drop across it and the tester will not light. If it does, either the fuse is no good or it's not making good contact in the socket.

23.11   When $f \to 0$ (that is, dc), the ratio is 1 (whatever low-frequency signals were in the input are for the most part in the output). This arrangement passes low frequencies on to the next circuit. When $f \to \infty$, the ratio approaches 0, little or no high-frequency voltage appears across C.

At $f \approx 0$, the capacitor is open circuited. At $f \approx \infty$, the capacitor is short circuited.

23.15   It's a one-transistor AM radio. The antenna feeds a tiny voltage (more or less, proportional to the length of the antenna wire) to the tuning circuit made up of the antenna coil and the variable capacitor. When tuned to resonate at the frequency of a particular station, current is maximum and the signal is passed on to the diode, which rectifies it (that is, it cuts off the negative portion). The transistor amplifies the positive signal so it will easily power the headphones, which respond to the envelope of the signal curve.

Answers to Multiple Choice Questions

   1. b   3. d   5. e   7. c   9. b   11. e   13. a   15. b

Solutions to Problems

23.5 $V_m = 1.414V$, since $V = 100$ V, $V_m = 141$ V.

23.15 $V = IX_C = 1/2\pi fC$; $C = 1/2\pi fV = (1.00$ A$)/2\pi(60$ Hz$)(120$ V$) = 22$ μF.

23.31   The voltmeter shows that the source is OK. If all functioned properly $R_e = 10$ Ω + 5.0 Ω + 50 Ω = 65 Ω; $I = V/R = (120$ V$)/(65$ $\Omega) = 1.8$ A and the fuse wouldn't blow. One of the resistors is malfunctioning, probably shorted internally. Even if both the 10 Ω and the 5.0 Ω were shorted the current would still be only 2.4 A so it must be the 50 Ω that's bad; $(120$ V$)/(15$ $\Omega) = 8.0$ A.

23.43   Using $y = mx + b$; $v(t) = \dfrac{2V_m}{T}t - V_m$; the average value of v^2 is $[v^2]_{av} = \dfrac{1}{T}\displaystyle\int_0^T v^2\, dt =$

$\dfrac{1}{T}\displaystyle\int_0^T \left(\dfrac{2V_m}{T}t - V_m\right)^2 dt = \dfrac{V_m^2}{3}$ and $V_{rms} = \sqrt{[v^2]_{av}} = \sqrt{V_m^2/3} = V_m/\sqrt{3}$.

23.61 $Z = V/I = 2307.7$ $\Omega = \sqrt{R^2 + X^2}$; $R = 1.7$ kΩ.

23.67 $L = 1/4\pi^2 f_0^2 C = 52.8$ mH.

23.71 $V_s = 30$ V; $V_p I_p = V_s I_s$; $I_s = 4.0$ A.

23.77 $Z = V/I = 100$ Ω; $Z = \sqrt{R^2 + X^2}$; $X = 97.98$ Ω; $X_C = 1/2\pi fC$; $C = 27$ μF.

23.83   efficiency = output/input = $(45$ kW$)/(45$ kW + 300 W + 500 W$) = 98.3\%$.

Answers to Discussion Questions

24.3   A hairdryer radiates radiowaves that can be picked up as noise in the picture on a nearby TV set.

24.7   The modern theory envisions light as both wave and particle; light energy is quantized but the propagation of those packets of energy is determined by its wave nature. Newton's version had a remarkably similar wave-particle structure: light was particulate but the particles were guided through space by wave patterns they set up in the aether.

24.9   The radiation is absorbed by the moist meat because it contains water. The plate is dry and stays cool. The water molecules in an ice cube cannot undergo rotational motion and will not absorb microwaves until some of the surface has melted.

24.13   A pulse can be a perfectly good wave. It's not obvious what is waving in a lightwave — it's certainly not a material aether. Later on, we'll talk about probability waves, but for the time being, let's say that the electromagnetic field itself waves.

24.15   The sphere is in equilibrium with its weight down, balanced by an upward force exerted by the beam. Radiant energy can therefore transfer momentum and exert pressure. A sail craft for space travel is quite possible.

Answers to Multiple Choice Questions

  1. c   3. b   5. a   7. b   9. a   11. d   13. c   15. d   17. b
  19. e   21. c

Solutions to Problems

24.7 $E = (20 \text{ V/m}) \cos 0 = 20 \text{ V/m}.$

24.15 $L = 1/4\pi^2 f_0^2 C = 1/4\pi^2 (100 \times 10^6 \text{ Hz})^2 (0.5 \times 10^{-12} \text{ F}) = 5 \ \mu\text{H}.$

24.29 $E = cB$, $E_0 = cB_0 = 2.0 \times 10^2 \text{ V/m}$; $B_0 = 6.7 \times 10^{-7} \text{ T}.$

24.35 $E_{max} = 1.00 \times 10^4 \text{ eV}$; $E = hf = hc/\lambda$; $\lambda = (4.136 \times 10^{-15} \text{ eV/Hz})(3.00 \times 10^8 \text{ m/s})/(1.00 \times 10^4 \text{ eV}) = 1.24 \times 10^{-10} \text{ m}.$

Answers to Discussion Questions

25.1   (a) It's mostly diffuse, although there is a little specular. (b) Flat paints have a diffuse, frequency-independent surface reflection that results in a white haze. (c) The index of water is between that of the fibers and the air so there is less diffuse white light reflected when it's wet. There will be a whitish haze over the dry painting.

25.7   It seems that the mirror is too far toward her middle for her to be looking at herself. She's looking at you which makes the picture strangely interesting.

25.9   The gentleman is standing where we are and her image is shifted too far to the right to be produced by a flat mirror parallel to the bar.

25.11   The resonance in the UV in part accounts for the effective absorption of UV by glass, which is essentially opaque to it. Only in great thicknesses will the weak absorption of red and blue produce a greenish tint.

25.13   (a) Red. (b) $C = W - R$, cyan ink "eats" red; hence, it will appear black. (c) It absorbs blue and produces more contrast between sky and cloud.

25.17   (a) $C = B + G$ — the filter absorbs B; hence, only G emerges. (b) $Y = R + G$ — the filter passes $R + B$ and absorbs G; hence, R is transmitted. (c) $(B + G) + (R + B) = (R + B + G) + B = W + B$, unsaturated blue. (d) $(R + G + B) - (R) - (G) = B$.

Answers to Multiple Choice Questions

   1. c   3. d   5. e   7. a   9. d   11. d   13. d   15. d   17. c
   19. e   21. d

Solutions to Problems

25.11   12.5 cm.

25.19   Draw a line from 0 perpendicular to the plane of the mirror and extend the lines $\overline{M_1M_2}$ and $\overline{C_1'C_2'}$ to meet at points A and B respectively. Triangles OAM_2 and OBC_2' are similar and so are OAM_1 and OBC_1' hence $H/d = h/(d + s_0)$ and so $H = hd/(d + s_0)$.

25.33 $n_a/n_w = d_A/d_R = 1/1.333 = 0.750 = 3/4$.

25.49   Light entering at glancing incidence is transmitted at the critical angle and those rays limit the cone of light reaching the fish; $\sin \theta_c = 1/1.333$; $\theta_c = 49°$ and the cone-angle is twice this or $98°$.

Answers to Discussion Questions

26.1   The focal length increases because the rays are not bent as strongly at the water-glass interface.

26.3   The focal length depends on the index of refraction and that depends on the wavelength.

26.9   The radius of curvative is ∞ and so is f. That means the object- and image-distances must have equal magnitudes. Thus, the magnification is +1.

26.13   The target is at one of the two focii of the hyperboloid and rays reflected from it appear to come from the other focus, $F_1(H)$, but this is also a focus, $F_1(E)$, of the ellipsoidal mirror. Rays appearing to come from one focus of the ellipsoid, after reflecting off it, converge to the other focus, $F_2(E)$, at the film plane.

26.17   The filament is located at one focus of the ellipsoid and the reflected light is made to converge at the other focus, which also corresponds to the front focal point of the lens combination.

Answers to Multiple Choice Questions

   1. b   3. b   5. d   7. b   9. a   11. e   13. b   15. d   17. a
   19. c   21. e

Solutions to Problems

26.7 $1/s_i = 1/f - 1/(100 \text{ cm}) = 0.006\ 67$; $s_i = 150$ cm.

26.19   For the standard observer $M_A = (0.254 \text{ m})\mathcal{D} = 8\times$; $\mathcal{D} = 31.5$ D and $f = 0.03$ m. The lens is 3 cm above the surface.

26.31   The total separation is 10.0 cm $= f_O + L + f_E = 1.0$ cm $+ L + 3.0$ cm; $L = 6.0$ cm.

26.39   We have $s_o = 0.10$ m and $s_i = 0.30$ m hence $f = s_o s_i/(s_o + s_i) = 0.075$ m. When $s_o = 0.025$ m, $1/s_i = 1/f - 1/s_o = 13.33 \text{ m}^{-1} - 40.00 \text{ m}^{-1}$, $s_i = -3.8$ cm. At first the image was real, inverted and magnified; after the jump it became virtual, erect, and magnified.

26.45 $M_T = y_i/y_o$ but since the two triangles are similar it follows that $M_T = -s_i/s_o$ where the image is again inverted. The greater the s_i the more the magnification.

26.53   The image will be inverted if it's to be real so the set must be upside down or else something more will be needed to flip the image; $M_T = -3 = -s_i/s_o$; $1/s_o + 1/3s_o = 1/(0.60 \text{ m})$; $s_o = 0.80$ m hence 0.80 m $+ 3(0.80 \text{ m}) = 3.2$ m.

26.61   (a) $1/(0.30 \text{ m}) + 1/(9.0 \text{ m}) = 1/f = -2/R$; $R = -0.58$ m. (b) $M_T = -s_i/s_o = (-9.0 \text{ m})/$ (0.30 m) $= -30$; the image is 1.5 m tall, real, and inverted.

26.65 $1/(500 \times 10^3 \text{ m}) + 1/s_i = -2/(-1.0 \text{ m})$; $s_i = 0.50$ m; $M_T = -(0.50 \text{ m})/(5 \times 10^5 \text{ m}) = 10^{-6}$. The image size is 2.0×10^{-6} m.

26.73   If s_i is the image-distance in the spherical mirror and β is the angle the image subtends, $\beta \approx y_{is}/(5.0 \text{ m} - s_i)$ where y_{is} is the image height and 5.0 m $- s_i$ is the distance from the observer to her image, s_i being negative. In the plane mirror $2\beta \approx (1.0 \text{ m})/(10 \text{ m})$, where the subject-image distance is 10 m; $\beta = 0.05$ rad. $M_T = y_{is}/(1.0 \text{ m}) = -s_i/(5.0 \text{ m})$ substituting this for y_{is} into the expression for β we get 0.05 rad $= [-s_i/(5.0 \text{ m})]/(5.0 \text{ m} - s_i)$ and $s_i = -1.666$ m. $1/(5.0$ m$) + 1/(-1.666 \text{ m}) = 1/f$; $f = -2.5$ m.

Answers to Discussion Questions

27.1   Yes. Monochromatic light can be imagined as the superposition of many waves with different polarization states but all of the same wavelength and all infinitely long. There cannot be any random phase changes because each component is infinitely long and constant in phase. So all will simply combine to form a constant polarization of some sort.

27.5   Since the index of refraction of benzene is around 1.5, the polarization angle will increase and the reflection again become visible.

27.7 $W = R + B + G$; red is strongly transmitted so the light passed is unsaturated red, and $B + G = C$ is reflected.

27.9   The speckle effect arises from the interference of light reflected from adjacent regions of the surface. The more coherent the illuminating light, the more apparent the phenomenon. With a laser the speckles almost fill the space in front of the illuminated surface.

27.11   For two holes on a horizontal line the pattern is as shown here. There will be a series of Young's fringes within the diffraction pattern of the individual apertures; namely, the Airy pattern.

27.13   Set the arms so that the optical path length for each is the same and then move one mirror until the contrast degrades so much that the fringes vanish; then $l_c = 2d$.

27.17   (a) 5; (b) 6; (c) 3; (d) 2; (e) 4; (f) 1; (g) 7.

Answers to Multiple Choice Questions

1. c   3. a   5. b   7. a   9. c   11. b   13. c   15. b   17. a

Solutions to Problems

27.7 $I = I_1 \cos^2\theta = (160 \text{ W/m}^2) \cos^2 60^\circ = \frac{1}{4} \, 160 \text{ W/m}^2 = 40 \text{ W/m}^2$.

27.11   The light from the first polarizer is at 10° and has an irradiance of $I_i \cos^2 30^\circ = 0.75 \, I_i$; this makes an angle of 60° with the next filter, hence $I_2 = (0.75I_i) \cos^2 60^\circ = 0.19I_i$.

27.23　　From Eq. (27.11), $d = m\lambda_0/2n_f = (500 \text{ nm})/2(1.36) = 1.84 \times 10^{-7}$ m.

27.31　　The center-line is a nodal line, i.e. the sound is a minimum along it because the sources are 180° out-of-phase. The first <u>maximum</u> will occur now when the path length difference is $\frac{1}{2}m\lambda$, hence $a \sin\theta = \frac{1}{2}m\lambda$, $ay_m/s = \frac{1}{2}m\lambda$ and $y_m = \frac{1}{2}m\lambda s/a = \frac{1}{2}mvs/af$ hence for $m = 1$, $y_1 = \frac{1}{2}(1) \times$ (346 m/s)(10.0 m)/(5.00 m)(1000 Hz) = 0.346 m.

27.43　　The two minima bounding the central maximum correspond to $m = \pm 1$ in Eq. (27.13); $\sin\theta_1 = \lambda/D = (461.9 \times 10^{-9} \text{ m})/(1.0 \times 10^{-4} \text{ m}) = 461.9 \times 10^{-5}$ and $\theta_1 = 0.26° = 4.6 \times 10^{-3}$ rad. The central band is twice this or 0.53°.

27.47　　From Eq. (27.8), $\Delta y = s\lambda/a$ and as the number of lines per centimeter doubles a is halved and so Δy, the space between orders, is doubled.

27.51　　From Eq. (27.15), $\theta_a = 1.22\lambda/D = 1.22(550 \times 10^{-9} \text{ m})/(5.08 \text{ m}) = 1.32 \times 10^{-7}$ rad = 7.57×10^{-6} degrees and multiplying by (60 min per degree)(60 s per min) yields 2.72×10^{-2} s.

27.63　　$\sin 20.0° = 1(500 \times 10^{-9} \text{ m})/a$, $a = 1.461\ 9 \times 10^{-6}$ m; $\sin 18.0° = 1\lambda/(1.461\ 9 \times 10^{-6}$ m), $\lambda = 451.75$ nm. Hence $n = (500 \text{ nm})/(451.75 \text{ nm}) = 1.11$.

Answers to Discussion Questions

28.1   (a) You would see yourself just behind where you are at the instant you spin around, moving backward in space and time until you plop down into the chair. The most distant image would take the longest time to reach you. (b) To see Lincoln, you need only fly away at a speed greater than c, overtake the wavefront corresponding to the desired event, pass it, and turn around. You can rush off at $v \approx 240c$, travel for a month or so, set up a telescope and watch yourself being born (assuming the great event occurred near a window with the shades up). (c) Of course none of this is possible since v must be less than c.

28.5   No. The scissors' contact point is more a mathematical notion than a physical one. There is no transport of energy from one point in space to another. Nothing actually goes outward from the pivot along the blades. The situation is the same for the overlap region of the two laser beams. It moves faster than c, but doesn't communicate anything. A long shadow can move faster than c.

28.7   The speed will remain constant at c, but there will be a Doppler shift lowering the frequency of the light. Since $E = hf$, a drop in frequency corresponds to a reduction in total energy. Because $E = pc$, the momentum must decrease as well. Photons have zero rest-energy and that doesn't change.

28.9   An electron moving along the wire, say, to the right, sees all the electrons in the first wire to be at rest and all the positive ions to be moving left — the electron sees the other wire move left. The observer electron in wire 2 sees a contraction of wire 1 and a resulting increase of positive charge density while the negative charge distribution is unchanged. The observer electron and all its fellows are attracted to the increased positive charge density — the wires attract.

28.11   As $v \rightarrow c$, $\gamma \rightarrow \infty$, and the total energy of the object becomes infinite. We can conclude that it takes an infinite amount of energy to bring a body up to light speed and therefore no object with mass can attain a speed of c.

28.15   Rewrite Eq. (28.7) as

$$\frac{v_{PO}}{c} = \frac{\left(\dfrac{v_{PO'}}{c}\right) + \left(\dfrac{v_{O'O}}{c}\right)}{1 + \left(\dfrac{v_{PO'}}{c}\right)\left(\dfrac{v_{O'O}}{c}\right)}$$

If $v_{PO'}/c = 1$ or $v_{O'O}/c = 1$, then $v_{PO}/c = 1$. When $v_{PO'}$ and $v_{O'O}$ are in the same direction, both + or both −, the denominator is > 1 and v_{PO} is less than its classical value, which keeps v_{PO} from exceeding c. When the signs of $v_{PO'}$ and $v_{O'O}$ are different and the motions are tending to cancel (as when someone runs to the rear of a moving train and stands still with respect to the platform), the denominator is < 1 and v_{PO} is larger than the classical value. That situation allows a light beam moving at $v_{PO'} = +c$ in a system moving at $v_{O'O} = -c$ to still be seen by someone at relative rest to be traveling at c.

28.17   We generalize the assumption and posit that <u>an object at one location in the Universe cannot affect another object that is a finite distance away instantaneously</u>. Then it follows that nothing can pass from one to another without a lapse of time and therefore <u>nothing can travel infinitely fast</u>. If there is an upper-limit to speed, and if it is measured to be different in different inertial systems, then it is possible to exceed the speed limit, and that contradicts the premise.

28.19   Envision an object interacting with another distant object via a stream of interaction particles traveling at a finite speed, call it c. If the first object now rushes ahead at a speed in excess of c, it can overtake the interaction particles and interact with itself, which is forbidden by the first postulate. Hence, c is the upper limit to speed.

Answers to Multiple Choice Questions

1. d   3. c   5. b   7. a   9. e   11. c   13. a   15. d   17. c

Solutions to Problems

28.7 $\Delta t_M = \Delta t_S / \sqrt{1 - v^2/c^2} = (60.0 \text{ s})/(0.0999) = 601$ s. The observer sees everything happening in slow motion and everybody laughs together or not at all.

28.15 $L_S = \gamma L_M = 1.25(400.0 \text{ m}) = 500$ m.

28.21   1.000 ft = 0.304 8 m, 1.000 yd = 3.000 ft = 0.914 4 m; $L_M/L_S = \sqrt{1 - v^2/c^2} = 0.914\,4$, $v^2/c^2 = 1 - 0.836\,1$, $v = 0.404\,8c$.

28.27 $L_M = L_S/\gamma = (1.000 \text{ m})/1.25 = 0.800$ m. $(0.800 \text{ m})/(0.600c) = 4.45$ ns.

28.29 $v_{PO} = (v_{PO'} + v_{O'O})/(1 + v_{PO'} v_{O'O}/c^2) = (c + v_{O'O})/(1 + c v_{O'O}/c^2) = c$.

28.33   Let y be perpendicular to the line of flight, then if the mast has a proper length of L, $L_y = L \sin 21.0^\circ = L_y' = 0.358\,4L$. The proper x-component of L is $L_x = L \cos 21.0^\circ = 0.933\,6L$ whereas $L_x' = 0.933\,6L/\gamma = 0.488\,8L$, $\tan \theta' = (0.358\,4L)/(0.488\,8L)$, $\theta' = 36.3^\circ$.

28.59 $p = \gamma mv = \gamma mc^2 v/c^2 = \gamma v(938.3 \text{ MeV})/c^2 = 100.0$ MeV/c, $\gamma v/c = 0.106\,57 = \beta\gamma = \beta/\sqrt{1 - \beta^2}$, $\beta^2/(1 - \beta^2) = 0.011\,357$, $\beta^2 = 0.011\,357/1.011\,357$, $\beta = 0.106\,0$, $v = 0.106\,0c$.

Answers to Discussion Questions

29.1   The NaCl disassociates into ions; namely, Na^+ and Cl^-. The chlorine carries its extra electron to the positive anode, where it gives it up and becomes neutral. The positive sodium ion, deficient by one electron (having given it to the chlorine), migrates to the cathode, where it picks up an electron and becomes neutral. The sodium atoms are not soluble and come out on the cathode as metallic sodium. The net result is that an electron has in effect traveled from cathode to anode that's the current. Had we started with a water solution, the water would have participated in the electrolysis and made things more complicated.

29.3   If it's finite in size, one might immediately ask, How is the charge distributed (that is, where on the electron is it)? If the electron has no parts, its charge has no parts either, and it becomes difficult to imagine any complicated distribution. Indeed if the charge were, say, spread over the surface of the electron (if it has a surface) what would keep the electron from blowing up under the Coulomb repulsion? Perhaps mass and charge are inseparable in the sense that whatever has mass has charge and vice versa.

29.7   The atoms in the first block are driven into oscillation in the zy-plane by the E-field, which is transverse to the x-axis. The beam traveling toward the second block is polarized. Atoms in the second block can only oscillate along the z-axis, and radiate in the xy-plane.

29.11   The wave number 109 677 cm^{-1} is the Series Limit of the Lyman Series just as 27 419 cm^{-1} is the Series Limit of the Balmer Series. All the levels correspond to the wave numbers of the successive Series Limits. It appears as if the atom, too, must have some sort of internal level structure such that a transition from one to another liberates light of a certain wavelength.

Answers to Multiple Choice Questions

1. d   3. b   5. b   7. c   9. b   11. a   13. b   15. c   17. a

Solutions to Problems

29.5   The net charge is (50.0 A)(5.00 min)(60 s/min) = 15 000 C; or 0.155 faraday; hence since valence is 1, 0.155 moles or 3.55 g.

29.13 $\sin \theta = m\lambda/2d$; $\sin \theta = 0.148\ 5$; $\theta = 8.54°$.

29.23   63.55 g is 1 gram-mole and contains 6.022×10^{23} atoms. Such a mass has a volume of (63.55 g)/(8.96 g/cm^3) = 7.092 6 cm^3 hence if each atom occupied a cube its volume is (7.092 6 cm^3)/(6.022 × 10^{23} atoms) = 1.177 8 × 10^{-23} cm^3 and has a side length of 2.28 × 10^{-10} m.

29.27 $e/m = E/B^2r = 1.758\ 8 \times 10^{11}$ C/kg; $B = 8.71 \times 10^{-4}$ T.

29.29 $KE = \gamma mc^2 - mc^2 = 1.00 \text{ MeV} = (0.511 \text{ MeV})(\gamma - 1); \gamma = 2.956\ 9; \beta = 0.941\ 1 = v/c;$
$v = 2.82 \times 10^8$ m/s.

29.35   From Eq. (29.4), $R = \dfrac{4(9.0 \times 10^9 \text{ N} \cdot \text{m}^2/\text{C}^2)(26)(1.60 \times 10^{-19} \text{ C})^2}{(6.62 \times 10^{-27} \text{ kg})(1.5 \times 10^7 \text{ m/s})^2} =$
1.6×10^{-14} m.

Answers to Discussion Questions

30.3   Classically, the re-emitted X-rays should come off in a spherical wave and both detectors should have picked up the radiation simultaneously. Since they didn't, the photon picture is upheld.

30.5 $Nhf = KE_{max} + \phi$, just increase the energy imparted to the electron by a factor of N. The maximum KE increases and so the stopping potential must increase, too. $KE_{max} = eV_s$. The work function is determined by the metal and does not change unless the metal itself is altered. $Nhf = \phi_0$, and so the threshold frequency $f_0 = \phi/Nh$ must decrease.

30.7   The light-quantum must provide an energy of at least E_a. Hence, the photon must have a minimum frequency $f_0 = E_a/h$, very much as in the photoelectric effect.

30.9   The initial momentum as seen in the center-of-mass system is zero, as is the final momentum. The electron is seen to be at rest after the collision. The initial energy is hf for the photon and $E = \gamma mc^2$ for the moving electron. The final energy is $E_0 = mc^2$ since the photon is gone and the electron is at rest (KE = 0). Conservation of Energy yields $hf + \gamma mc^2 = mc^2$, which implies $m > \gamma m$.

INITIAL

$p = h/\lambda$

$p = mv$

FINAL

$v_f = 0$

30.11   Atoms are pumped up to the 4th level from the ground state, emptying the latter. They immediately drop to the 3rd level, which is metastable and thus unlikely to experience many spontaneous emissions. There is then a population inversion between the 3rd and the 2nd levels and a laser transition occurs from the former down to the latter. Rapid decay from the 2nd to the ground state ensures that the inversion will be maintained, an improvement over the 3-level system.

30.13   Provided the atom was being illuminated by the proper light, it was absorbing and re-emitting photons almost continuously as it bounced up and down, into and from, the lower-excited state. The key was that the higher level was metastable. An atom can only be in one excited state at a time and so whenever the atom was kicked into the higher state, it could not absorb and re-emit via the lower one. The bright light from the atom blinked off whenever it went into the metastable state, and it blinked back on as soon as the atom dropped out of the metastable state.

Answers to Multiple Choice Questions

1. c   3. a   5. b   7. b   9. c   11. d   13. b   15. c   17. b

Solutions to Problems

30.3   From Eq. (30.4) $\lambda_{max}T = 0.002\ 898$ m·K; $T = 306$ K; $\lambda_{max} = 9.4\ \mu$m i.e., infrared.

30.15   Using Table 30.2, $KE_{max} = hf - \phi = hc/\lambda - \phi = 1.89$ eV.

30.21   60 keV − 25 keV = 35 keV.

30.33 $KE_{max} = eV_S = (1.60 \times 10^{-19}$ C$)(1.250$ V$) = 2.00 \times 10^{-19}$ J. And $v = \sqrt{2(KE_{max})/m_e} = 0.663 \times 10^6$ m/s.

30.37 $hf = \Delta(KE) = \frac{1}{2}m_e(v_i^2 - v_f^2) = 3.416 \times 10^{-15}$ J; $f = 5.16 \times 10^{18}$ Hz. $\lambda = c/f = 5.82 \times 10^{-11}$ m.

30.45 $T = 2\pi r/v$; $1/T = f = v/2\pi r$; $v_n = nh/2\pi m_e r_n$; $f_n = nh/(2\pi r_n)^2 m_e$ using Eq. (30.17); $f_n = m_e k^2 Z^2 e^4/2\pi n^3 \hbar^3$

Answers to Discussion Questions

31.1   Since $\lambda = 1.25/\sqrt{V}$ nm, the wavelength can be quite small. For example, at 100 kV, $\lambda = 0.004$ nm. Although electrons can be focused easily, X-rays of comparable wavelength are very difficult to focus (though progress is being made in that endeavor). Electrons (in vacuum to prevent scattering) pass through a thin slice of the sample where they are scattered off in a pattern that corresponds to the information. This beam is collected by the objective, which forms a magnified intermediate image. The electrons from that image are collected and the image further enlarged by the projection lens. The product of these two magnifications can be an enlargement of 200 000× or so.

31.5   According to classical theory, a particle passes through one hole, a wave through both holes. If either coil records the presence of an electron, we are dealing with the particlelike manifestation of the electron and the wavelike behavior will vanish. Thus, the theory maintains that the interference pattern will vanish as soon as either coil picks up the transit of an electron. Presumably, the induced current will produce an induced magnetic field, which will destroy the interference.

31.7   If the particle is confined, there will be a nonzero Δx and therefore a nonzero Δp, which is the minimum p the particle must have and it corresponds to a minimum $KE = p^2/2m$. The atoms in a box at absolute zero must themselves still be moving around with a zero-point energy.

31.9   A range of frequencies and wavelengths is needed to synthesize the packet and, hence, a range of frequency (energy) and wavelength (momentum) is always present, thus constituting an uncertainty in E and p for the particle. To make $\Delta p = 0$, we must use a monochromatic wave, one wavelength, one p, and no uncertainty. But where will the particle be if the wave is a perfect sine wave? Anywhere, and so $\Delta x = \infty$. If the packet shrinks so $\Delta x \rightarrow 0$, the number of sine waves needed to make the packet increases to infinity, and $\Delta p \rightarrow \infty$.

31.11   The peaks occur at the Alkali Metals, which have a single outer electron in an unfilled shell. This electron is shielded from the nucleus by all the filled shells below it, and so is held weakly. If the nuclear charge is +Ze, the electron "sees" a charge of just +e. Accordingly, the electron cloud is relatively large. Similarly, boron, aluminum, and gallium atoms have three outer electrons that see a nuclear charge of +3e and are strongly bound in small orbits. Thus, the curve rises and falls.

Answers to Multiple Choice Questions

   1. b   3.   5. a   7. d   9. d   11. c   13. d   15. a

Solutions to Problems

31.3 $\lambda = h/p = h/mv = 10h/m_e c = 2.4 \times 10^{-11}$ m.

31.7 $eV = \frac{1}{2}mv^2 = \frac{1}{2}m(p/m)^2 = \frac{1}{2}p^2/m = \frac{1}{2}(h/\lambda)^2/m$; $V = (h/\lambda)^2/2em = 0.15$ kV.

31.11 K has 2; L has 8; M has 18; N has 32. The general formula for the number of electrons is $2[1 + 3 + ... + (2n - 1)]$.

31.19 $\Delta t = 4.4 \times 10^{-24}$ s; $\Delta E = \frac{1}{2}\hbar/\Delta t = (3.29 \times 10^{-16}$ eV·s$)/(4.4 \times 10^{-24}$ s$) = 75$ MeV; $\Delta E/E = 9.8\%$.

31.35 $\Delta p/p = 1/1000$; $\Delta p = 10^{-3}p = 10^{-3}mv$; $\Delta x \geq \frac{1}{2}\hbar/\Delta p = \frac{1}{2}\hbar/(10^{-3}\,mv) = 2.64 \times 10^{-29}$ m.

Answers to Discussion Questions

32.3 The atom-bomb trigger generates temperatures in excess of 100 million K. The lithium is blasted by neutrons and converted into tritium, which then combines via fusion with deuterium, liberating large amounts of energy. The fast neutrons ($\approx$ 14.1 MeV) then fission the surrounding U-238 "blanket" which can be as large as can be delivered since there's no concern about critical mass. Tritium can be separated from sea water, but it's usually produced by putting lithium-6 in a fission reactor.

32.5 The surface of the pellet is vaporized, driving the remainder violently inward. The inner core compresses to tremendous densities (1000 times that of water). Rising to temperatures in excess of 100 million degrees, the thermonuclear processes begin and there is a mini-H-bomb explosion (about the equivalent of 50 lb of high explosives). A constant drop and blast is the ultimate goal. $^{3}_{1}H + {}^{2}_{1}H \rightarrow {}^{4}_{2}He + {}^{1}_{0}n + 17.6$ MeV

32.7 The half-life is the time it takes the sample to decay to half its original value. The nuclear mean lifetime is $1/0.693 = 1.44$ times longer than the half-life — the average radionuclide will live that long. The half-life of an American human is 68 years; half the people born 68 years before will be dead. But the mean human life is certainly not $1.44(68\ y) = 98\ y$. The nuclei don't age, we do. Our chances of surviving two half-lifes (136 y) are zero. By comparison, some few nuclei will go on for 100 half-lives and more.

Answers to Multiple Choice Questions

1. d   3. b   5. b   7. d   9. b   11. a   13. d   15. b   17. a
19. e

Solutions to Problems

32.1 111.

32.11 $(1.542\ 748 \times 10^{-25}$ kg$)/(1.660\ 540 \times 10^{-27}$ kg/u$) = 92.906\ 38$ u.

32.15 $R = 3.6$ fm $= (1.2$ fm$)A^{1/3}$; $A = 3^3 = 27$; the element is aluminum.

32.23 $50.7\%(78.918\ 336$ u$) + X\%(80.916\ 289$ u$) = 100\%(79.909)$; $X\% = 49.3\%$ and $49.3\% + 50.7\% = 100\%$; there are no other stable or very long lived isotopes.

32.29 $\Delta m = 26(1.007\ 825$ u$) + 28(1.008\ 665$ u$) - 53.939\ 613$ u $= 0.506\ 457$ u. $E_B/A = 8.736$ MeV.

32.33 Mass initially - mass finally; $(232.037\ 13$ u$) - (228.028\ 73$ u$) - (4.002\ 603$ u$) = 0.005\ 80$ u $\rightarrow 5.40$ MeV.

32.39   Mass before – mass after = Q, 8.023 828 u – 8.005 206 u = 0.018 622 u → 17.346 MeV.

32.53   (9.012 182 u) + (4.002 603 u) – (12.000 000 u) – (1.008 665 u) = 0.006 120 u = 5.700 74 MeV.

32.57   We need $R = N\lambda$; to find N, 1 mg → $(10^{-6}$ kg)/(222 kg/kmole) = 4.505 × 10^{-9} kmole → 2.713 × 10^{18} atoms = N; $R = 5.7 × 10^{12}$ Bq.

Answers to Discussion Questions

33.1   The collider smashes a beam of electrons head-on into a beam of positrons with a combined energy of $\approx$ 100 GeV. This quantity is more than enough to create Z^0 bosons — this is a so-called Z^0 factory, albeit a rather feeble one. The two electron pulses are accelerated (up to 1 GeV), condensed, and injected into the linac (linear accelerator). They are joined by a previously processed positron bunch, which is further accelerated, along with the leading electron bunch, to the end of the linac. The trailing electron bunch is diverted to the side, where it produces a shower of positrons that is sent back to the start for processing so that it can join the next pulse of electrons. Meanwhile, the two high-energy bunches are turned around by magnets at the end of the linac and they collide in the detector.

The drift-tube linac is a succession of tubular conductors that have voltages applied to them just at the right moment so the electrons (or positrons) are accelerated across each gap. The particles see no E-field inside the conductors and drift along them. As they move faster, the cylinders are longer, so the generator frequency can be constant.

33.3   For the most part, the basis for the distinction between matter and antimatter arises by comparison to the stuff of our ordinary existence. If a baryon (for example, Λ) decays to ordinary matter (Λ decays into a neutron), we call it matter. If it decays to antimatter ($\bar{\Lambda}$ decays into an antineutron), we call it antimatter. But the mesons are half-and-half (quark-antiquark), so we can expect ambiguity. There is no way to determine, and no need to determine, which pion is the antipion.

33.7   A negatively charged pion entered the chamber leaving a clear slightly curved track, which means that it had a good deal of linear momentum ($R \propto p$). It struck a proton and created two nonionizing neutral particles (K^0 and Λ^0) that left no tracks. Being unstable, they decayed via the weak force (thus living long enough to traverse a substantial distance). The entire process was discussed in Questions 5 and 6.

Answers to Multiple Choice Questions

   1. d   3. e   5. c   7. c   9. a   11. a   13. d   15. a   17. c
   19. b

Solutions to Problems

33.7   No. Electron-lepton number is not conserved.

33.11   No. It conserves charge, lepton number, spin and with a weak decay we needn't worry about strangeness.

33.15   (1115.6 MeV) - (139.6 MeV) - (938.3 MeV) = 37.7 MeV. It needn't conserve strangeness.

33.27   (139.6 MeV) + (938.3 MeV) - (497.7 MeV) - (1115.6 MeV) = -535.4 MeV. No, a considerable amount of energy has to be present as KE initially.

33.31   As a meson it's $q\bar{q}$; must have charm so one quark is c; that gives a charge of +2/3 so the next quark must have $Q = -2/3$, $S = 0$, $C = 0$ and that makes it $\bar{u}$: $D^0 = c\bar{u}$.

33.35 $d\bar{s} \rightarrow u\bar{d} + d\bar{u}$; the $\bar{s}$ decays via the weak force into a $\bar{d}$ plus energy which creates a u-$\bar{u}$ pair, yielding the original d plus $\bar{u}$ plus $u\bar{d}$.

33.39   Using Yukawa's result $R \approx h/4\pi mc \approx hc/4\pi mc^2$; $mc^2 \approx hc/4\pi R \approx (1.239\ 8 \times 10^{-6}\ \text{eV·m})/4\pi(10^{-31}\ \text{m}) \approx 10 \times 10^{23}\ \text{eV} \approx 10^{15}\ \text{GeV}$.

Need help solving the text's problems?

Introducing *Brooks/Cole Exerciser (BCX) v 2.0 for DOS, Windows, and Macintosh*

BCX is an easy-to-use software program that helps you solve the types of problems that appear in Hecht's *Physics: Calculus*

Features

- Each section in *BCX* begins with a problem from the text. *BCX* walks you through solving this problem.
- For each *BCX* problem, you are asked to indicate the correct solution. Choosing a correct answer prompts another problem, but choosing an incorrect answer repeats the question and offers a hint.
- A second wrong answer prompts *BCX* to provide the complete solution and an illustration of the concepts behind the problem.
- *BCX*'s record-keeping capabilities allow you to track how you are doing and to rework problems to improve your score.
- You can print your score or view it on the screen.
- The "Bookmark" feature allows you to quit the program halfway through a section and to return easily to that exact point.

Price: $20.25

DOS: ISBN: 0-534-33987-5. **Mac**: ISBN: 0-534-34171-3. **Windows**: ISBN: 0-534-34168-3.

System Requirements

DOS: 80x86 or later. DOS 3.2 or later. EGA, VGA, or SVGA monitor. 400 KB free RAM. 2.5 MB available on hard drive.

Macintosh: Mac SE or later, but not native PowerMac. System 6.0.5 or later. Monochrome or color monitor. 1 MB available on hard drive.

Windows: 80x86 or later. Windows 3.1 in Enhanced Mode or Windows 95. VGA or better graphics card. 1 MB free memory. 2.5 MB available on hard disk.

To order, please use the attached coupon or call 1-800-354-9706.

FOLD HERE

NO POSTAGE
NECESSARY
IF MAILED
IN THE
UNITED STATES

BUSINESS REPLY MAIL

FIRST CLASS   PERMIT NO. 358   PACIFIC GROVE, CA

POSTAGE WILL BE PAID BY ADDRESSEE

ATT: MARKETING

Brooks/Cole Publishing Company
511 Forest Lodge Road
Pacific Grove, California 93950-9968

FOLD HERE

Master the calculus problems in Hecht's
Physics: Calculus
with this step-by-step workbook!

Solving the calculus problems in Hecht's **Physics: Calculus** is easier using Zvonimir Hlousek and Regina L. Neiman's **Calculus Problem Workbook.**

Need help in approaching and setting up calculus-based problems?
You'll appreciate the examples (with detailed solutions) that stress the basics of each chapter.

Want more practice in problem solving?
You'll want to use the additional exercises. Approximately 60% are easy problems that you can use to check how comfortable you are with the material. The other 40% are more challenging. Answers to all appear in the back of the manual.
ISBN: 0-534-32399-5.

To order your copy, please fill out and return the order form below. We'll rush you a copy just as soon as we hear from you.

ORDER FORM

To order, simply fill out this coupon and return it to Brooks/Cole along with your check, money order, or credit card information.

Yes! I would like to order *Calculus Problem Workbook for* Hecht's *Physics: Calculus* by *Zvonimir Hlousek and Regina L. Neiman.*

_____ $16.00. (ISBN: 0-534-32399-5).

(Residents of: AL, AZ, CA, CT, CO, FL, GA, IL, IN, KS, KY, LA,MA, MD, MI, MN, MO, NC, NY, OH, PA, RI,SC, TN, TX, UT, VA, WA, WI must add appropriatstate sales tax.)

Subtotal_____
Tax _____
Handling _____2.00_____
Total _____

Payment Options
____Check or money order enclosed.
or bill my ___ VISA ___ MasterCard ___American Express / Card Number: _____
Expiration Date: _____ Signature: _____
Note: Credit card billing and shipping addresses must be the same.

Please ship my order to: (please print)
Name _____ Street Address _____
City _____ State _____ Zip+4 _____Telephone (_____) _____

Mail to:Brooks/Cole Publishing Company, Dept. Hechtinbook, 511 Forest Lodge Road, Pacific Grove, CA 93950-5098
Phone: (408) 373-0728 or Fax: (408) 375-6414
Prices subject to change without notice.

SECURE WITH TAPE